Adaptive Control of Mechanical Manipulators

John J. Craig
Silma, Inc.

ADDISON-WESLEY PUBLISHING COMPANY
Reading, Massachusetts • Menlo Park, California • New York
Don Mills, Ontario • Wokingham, England • Amsterdam • Bonn
Sydney • Singapore • Tokyo • Madrid • Bogotá • Santiago • San Juan

Library of Congress Cataloging-in-Publication Data

Craig, John J., 1955–
  Adaptive control of mechanical manipulators.

  Bibliography: p.
  Includes index.
  1. Manipulators (Mechanisms)—Automatic control.
2. Adaptive control systems. 3. Robots, Industrial.
I. Title.
TJ211.C66 1988 629.8'92 86-22158
ISBN 0-201-10490-3

ABCDEFGHIJ-AL-8987

PREFACE

This book is concerned with one of the most interesting and challenging problems in controlling a robot manipulator—the use of adaptive control algorithms. The idea of adaptive or learning control for mechanical manipulators has certainly come to the mind of anyone who has considered the problem of controlling such a complex system. It is a compelling notion that one might build a system that can learn and improve its performance as it operates. This desire has led many researchers to investigate the application of adaptive strategies to general control problems, and recently, to the problem of robot control. The problem is not an easy one, and is not yet completely solved.

This book examines the problem in the nonlinear domain in which the robot control problem is set, rather than relying on the existing theory for adaptive control of linear systems. Along the way, the theory for linear systems is reviewed, previous research in adaptive control of robot manipulators is reviewed, and manipulator dynamics and nonadaptive control are presented.

New results presented in the text include a robustness result for the non-adaptive model-based robot controller, and an algorithm for the learning control of manipulators. The core of the book is the presentation of a new non-linear adaptive controller for mechanical manipulators that is rigorously proven globally stable.

This book is appropriate for mathematicians and engineers with an interest in manipulator dynamics and control. In particular, those interested in the topics of robustness, adaptation, and learning in the context of nonlinear systems such as robot manipulators may find the text interesting.

Acknowledgments

I wish to thank many people who contributed in various ways to this book. My principal thesis advisor, Prof. Bernard Roth, deserves special thanks for this technical guidance and wisdom. It was my pleasure to have Prof. Roth as an advisor and friend at Stanford. Many results in Chapter 5 are joint conclusions with my colleagues at the University of California at Berkeley, Ping Hsu and Prof. S. Shankar Sastry. At Stanford, I benefited from discussions with Dr. Oussama Khatib, Prof. Gene Franklin, and Prof. Stephen Boyd. I am thankful to Greg Martin who worked on the simulations appearing in Chapter 5. Jim Maples supplied many hours of help with the experiments at Adept Technology, Inc.

The help and enthusiasm of the following is gratefully acknowledged: Brian Armstrong, Joel Burdick, Madhusudan Raghaven, Dr. Ron Goldman, Dr. Jeff Kerr, Dr. Dave Marimont, John Hake, Dr. Bill Hamilton, Brian Carlisle, Prof. Bernard Widrow, and Prof. Gordon Kino.

I am grateful to the Systems Development Foundation for their support through contracts with Prof. Roth. Several years of support were provided by Prof. Tom Binford. Silma, Inc., provided support and was understanding in scheduling my time carefully during completion of the book. Adept Technology, Inc., provided the use of their facilities for the experiments of Chapter 5.

This book is essentially a reprint of my Ph.D. dissertation, but I am thankful to the editors and staff at Addison-Wesley for their input, which led to many manuscript corrections.

For moral and spiritual support, I must single out Monique Craig, who has inspired and pushed me. I also thank my parents, James and Harriet Craig, and Tom and Linda Craig, Connie Craig, Chris Goad, Donalda Speight, Tim Turner, Paul and Avi Munro, Al Barr, Warren Davis, Joe Levy, Dave Baras, Ned Kirchner, Peter Hochschild, Glenn Howland, and others. Finally, special thanks to those smart enough to transcend all this: Yoda, Zorro, Babar, and Kaza.

Palo Alto, California   J.C.

CONTENTS

Chapter 1

INTRODUCTION

1.1 Introduction

Present-day industrial robots operate with very simple controllers, which do not yet take full advantage of the inexpensive computer power that has become available. The result is that these fairly expensive mechanisms are not being utilized to their full potential in terms of the speed and precision of their movements. With a more powerful control computer it is possible to use a dynamic model of the manipulator as the heart of a sophisticated control algorithm. This dynamic model allows the control algorithm to "know" how to control the manipulator's actuators in order to compensate for the complicated effects of inertial, centrifugal, Coriolis, gravity, and friction forces when the robot is in motion. The result is that the manipulator can be made to follow a desired trajectory through space with smaller tracking errors, or perhaps move faster while maintaining good tracking.

There are two reasons why such sophisticated control algorithms have not found use outside of research laboratories. The first is the economics of supplying sufficient computing power to the robot controller. Recently this problem has diminished greatly, and will continue to do so. The second and more serious problem is that of imprecise dynamic models. Developing a correct dynamic model (in the form of a set of coupled differential equations) for a multidegree-of-freedom manipulator is a difficult task. Recent work has made developing the structure of these equations more or less straightforward for the cases where the links are modeled as rigid bodies. However, a problem that remains is that of unknown parameters that appear in the model, and also of effects such as friction and flexibilities, which are left out of the model

1

formulation. Not only are parameters unknown or only poorly known, but they also may be subject to change as the manipulator goes about its tasks.

This book addresses the control of mechanical manipulators in cases where the physical models that describe the manipulators are not well known. Incorrectness or uncertainty in a dynamic model can be split into two portions. *Structured uncertainty* is what we will call the case of a correct structural model with all uncertainty due to incorrect parameter values. That is, there exists a correct (but unknown) set of values for the parameters such that the model will match the actual system. *Unstructured uncertainty* is the name given to unmodeled effects, some of which may be state-dependent and some of which are external disturbances. Unstructured uncertainty arises from sources not considered by the designer, or those that are too complex to model.

Much of this book addresses the case where modeling error is largely due to structured uncertainty. However, the reality of external disturbances is considered throughout, and in Chapter 6 a special learning algorithm is developed specifically for the case of unmodeled dynamic effects that lack a parametric model.

1.2 Historical Summary

In the past several years a great number of papers have been published about various aspects of robotics. Even confining ourselves only to those papers dealing with the control problem for mechanical manipulators, the volume of published work makes a concise review difficult. We will therefore mention only early work, and only in the areas of kinematics, path generation, dynamics, and control. We will neglect many important areas such as sensors, robot programming languages, locomotion, etc. Although we believe that we have cited the most important contributors, it is possible that we are unaware of some work.

Today's industrial robots have their roots in numerically controlled (NC) machines and the early master–slave teleoperators used in the nuclear industry. In 1947 work started at Argonne National Laboratory on master–slave systems. Originally these were simply mechanical linkages; later electrically and hydraulically powered systems were developed, some with "force-reflecting" capability [1].

Based on George Devol's ideas, Unimation Inc. developed the first industrial robot in 1959, and installed the first robot in a U.S. factory in 1961 [2]. In 1961 Ernst [3] developed a computer-controlled mechanical hand with tactile sensors, called "MH-1," which was coupled with an Argonne National Laboratory manipulator and a computer. It was capable of stacking blocks under computer control.

At Stanford University an early laboratory was established in 1965 by John McCarthy and others [4]. One of the first six degree-of-freedom, electric, computer-controlled manipulators was designed and built by Scheinman in 1969 [5], and became known as the Stanford Scheinman arm.

Early work in robotics was largely concerned with the basic problems of representing spatial information [6], and the manipulator's kinematic equations and their solution [7, 8]. Another early focus of research was in generating trajectories and controlling the manipulator to move along them [9-13].

The first application of dynamic analysis to the particular problem of a multidegree-of-freedom mechanical manipulator seems to have been by Kahn and Roth [12], based on Uicker's work [14] on linkages. This early work was not particularly concerned with efficiency and resulted in a computational algorithm that was $O(n^4)$ in complexity, where n is the number of manipulator joints. Renaud [15] and Liegois et al. [16] made early contributions concerning formulating the mass-distribution descriptions of the links. While studying the modeling of human limbs, Stepanenko and Vukobratovic [17] began investigating a "Newton–Euler" approach to dynamics instead of the somewhat more traditional Lagrangian approach. This work was revised for efficiency by Orin et al. [18] in an application to the legs of walking robots. Orin's group improved the efficiency somewhat by writing the forces and moments in the local link reference frames instead of the inertial frame. They also noticed the sequential nature of calculations from one link to the next, and speculated that an efficient recursive formulation might exist. Armstrong [19] and Luh, Walker, and Paul [20] paid close attention to details of efficiency and published an algorithm that is $O(n)$ in complexity. This was accomplished by setting up the calculations in an iterative (or recursive) nature and by expressing the velocities and accelerations of the links in the local link frames. Hollerbach [21] and Silver [22] further explored various computational algorithms. Hollerbach and Sahar [23] showed that for certain specialized geometries the complexity of the algorithm would further reduce. Finally, several authors have published articles showing that for any given manipulator customized closed-form dynamics are more efficient than even the best of the general schemes [24-29].

The net effect of the developments in computing the dynamic model of a manipulator, coupled with the increasing power of computers, was that the model could be computed sufficiently quickly for use in real-time control. The use of the nonlinear dynamic model of a manipulator in a control algorithm apparently has its roots in the work of several researchers. Perhaps the earliest is the work of Freund [30, 31], in which he uses tools from Lie Algebra to discuss the decoupling and linearizing of nonlinear systems. The work of Bejczy [32], Lewis [33], and Markiewicz [34], seems to be responsible for the term "computed torque method" by which the general approach is sometimes known. Other early work was done by Zabala-Iturralde [35], Khatib et al.[36], and Liegois et al. [37]. A closely related approach was given by Luh, Walker, and Paul in [38].

Later, there were many different papers published on various approaches to manipulator control. Of these methods, we will confine our discussion to adaptive techniques. A complete review of adaptive control applied to manipulators is defered until Chapter 4, but the earliest work seems to have been done by Timofeyev and Ekalo [133], Dubowsky and DesForges [83], and Horowitz and Tomizuka [90].

1.3 Contributions of This Book

This book addresses the control of mechanical manipulators in cases where the physical models that describe the manipulators are not well known. We adopt the view that the nonlinear, model-based method (or "computed torque method") of manipulator control is, in theory, a good approach to manipulator control. We then investigate methods of compensating for the fact that a perfect dynamic model is never available.

The contributions of this book are in three areas:
(1) The robustness of the model-based servo in the presence of poorly known parameters is investigated. A sufficient condition for stability of the overall system in the presence of parameter errors is developed.

(2) A parameter-adaptive control scheme is developed in the form of a set of adaptive laws that can be added to the nonlinear model-based controller.

The scheme is unique in that it is designed specifically for this model-based controller, and is rigorously proven stable in the full nonlinear setting.

(3) A learning control scheme is developed that can be added to the model-based controller in order to "learn" compensation for friction and other effects that are difficult to model parametrically.

1.4 Preview of the Book

In Chapter 2 we examine some properties of the dynamics of manipulators and then examine a couple of control stategies that are in use or have been proposed for their control. One of these methods, the so-called computed torque servo, or nonlinear model-based control scheme, will be the focus of our attention. This scheme represents an excellent way to use a dynamic model of a manipulator (if it were perfectly known) in a controller.

In Chapter 3 we analyze some robustness properties of the nonlinear model-based control scheme. The basic question is When the parameters appearing in the model are not well known, does the control scheme still perform well?

In Chapter 4 we review the notion of adaptive control as a methodology for compensating for unknown or loosely known parameters. We review the adaptive control stategies for mechanical manipulators that have been proposed by other researchers.

In Chapter 5 we derive a new adaptive control scheme for manipulators and discuss its properties. The scheme is novel in that no assumptions of plant linearity are made in the development of the adaptation laws or in the stability proof. The scheme can be viewed as an extension of the existing theory of adaptive control for linear systems to a class of nonlinear systems that includes rigid-body models of manipulators. We also consider the questions of persistent excitation and robustness to bounded disturbances.

In Chapter 6 we present a method for learning control of manipulators. This scheme may be used on its own, or in conjunction with the adaptive scheme of Chapter 5. Dynamic effects for which a parametric model are unavailable may be handled by this scheme.

In Chapter 7 we present some conclusions.

In Appendix A there is a brief introduction to norms and normed spaces. Appendix B presents a brief introduction to Lyapunov stability theory. In Appendix C the notion of a strictly positive real transfer function is introduced.

Throughout the book, theories are illustrated by results from simulations, and, in Chapter 5, by results from actual experiments with an industrial manipulator.

Chapter 2

CONTROL OF
MECHANICAL MANIPULATORS

2.1 Introduction

In this chapter we present the basics of the trajectory-control problem of mechanical manipulators. First we state our underlying assumptions, which will hold throughout the book. These help to outline the scope of the book by stating what is and what is not assumed to be true of the manipulator system. Next, the dynamic equations that describe the motion of a manipulator are presented, along with some notes on the inherent structure of these equations. We look briefly at one simple method before introducing the scheme on which we will focus, the so-called *computed torque* method of manipulator control.

Here we state some assumptions that will hold throughout this book.

(1) Manipulators are modeled as jointed rigid bodies. Link flexibility will not be addressed as such. Often the analysis will provide for a bounded "disturbance torque," T_d, acting at the joints. This may in some circumstances represent some of the torques attributable to flexibility, but not in any rigorous sense, because T_d will be assumed to be uncorrelated with the manipulator state, an assumption not necessarily true for torques caused by flexible bending modes in the linkages.

(2) Continuous-time analysis is employed. Discrete-time sampling and control effects will not be addressed in the analysis. On the other hand, some numerical simulations do include discrete sampling and control effects. Also, actual experiments with a physical manipulator will be

presented, for which, obviously, the control was performed by a computer in discrete time. Hence, analysis is performed in continous time, and simulations and experiments are used to verify empirically that the theory is implementable.

(3) Manipulators under discussion will be assumed to be open kinematic chains with revolute joints. With a very small amount of effort, all analyses performed can be extended to open chains with mixed revolute and prismatic joints, and with somewhat more effort, results can be extended to systems that contain closed kinematic chains.

(4) The desired manipulator joint trajectories will be assumed to be known, including first and second derivatives. Such trajectories could be computed by any of several well-known methods. We will assume that the desired trajectories are smooth, meaning that desired angular velocities and angular accelerations are bounded.

These assumptions will be in effect in the sequel unless otherwise stated.

2.2 Structure of the Manipulator Dynamic Equations

This section discusses the structure of the dynamic equations of motion for a mechanical manipulator. Results from this section will be useful throughout the remainder of the book.

The manipulator is modeled as a set of n moving rigid bodies connected in a serial chain with one end fixed to the ground and the other end free (Figure 2.1). The bodies are jointed together with revolute joints, there is a torque actuator and friction acting at each joint. The vector equation of motion of such a device can be written in the form [42]

$$T = M(\Theta)\ddot{\Theta} + V(\Theta, \dot{\Theta}) + F(\dot{\Theta}) + G(\Theta) + T_d, \qquad (2.1)$$

where T is the $n \times 1$ vector of joint torques supplied by the actuators, and Θ is the $n \times 1$ vector of joint positions, with $\Theta = [\theta_1, \theta_2, \ldots, \theta_n]^T$. The matrix, $M(\Theta)$, is an $n \times n$ matrix, sometimes called the manipulator mass matrix. The vector $V(\Theta, \dot{\Theta})$ represents torques arising from centrifugal and Coriolis

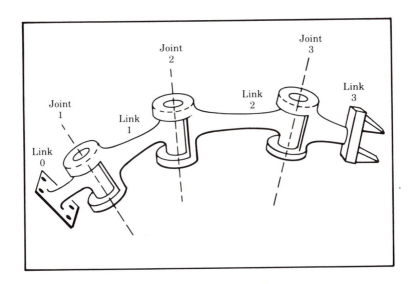

Figure 2.1 *An Articulated Chain of Rigid Bodies*

forces. The vector $F(\dot{\Theta})$ represents torques due to friction acting at the manipulator joints. The vector $G(\Theta)$ represents torques due to gravity, and T_d is a vector of unknown signals due to unmodeled dynamics and external disturbances.

Also, we will sometimes write the dynamics in the more compact form

$$T = M(\Theta)\ddot{\Theta} + Q(\Theta, \dot{\Theta}) + T_d, \tag{2.2}$$

where the vector $Q(\Theta, \dot{\Theta})$ represents torques arising from centrifugal, Coriolis, gravity, and friction forces.

It will be convenient from time to time to write the vector of velocity terms in the matrix–vector product form:

$$V(\Theta, \dot{\Theta}) = V_m(\Theta, \dot{\Theta})\dot{\Theta}, \tag{2.3}$$

where the subscript m stands for "matrix."

The jth element of (2.2) can be written in the sum-of-products form

$$\tau_j = \sum_{i=1}^{u_j} m_{ji} f_{ji}(\Theta, \ddot{\Theta}) + \sum_{i=1}^{v_j} q_{ji} g_{ji}(\Theta, \dot{\Theta}) + \tau_{dj}, \tag{2.4}$$

where the m_{ji} and q_{ji} are parameters formed by products of such physical quantities as link masses, link inertia tensor elements, lengths (e.g., distance to a center of mass from a joint), friction coefficients, and the gravitational acceleration constant. The $f_{ji}(\Theta, \ddot{\Theta})$ and the $g_{ji}(\Theta, \dot{\Theta})$ are functions that embody the dynamic structure of the motion geometry of the manipulator.

In this book we generally assume that the structure of these parameters and functions are known, but the numerical values of some or all of the parameters m_{ji} and q_{ji} are poorly known. However, we will assume that bounds on the parameter values are known.

2.2.1 The Manipulator Mass Matrix, $M(\Theta)$

Intuitively, it should be possible to write the kinetic energy of a mechanism like a manipulator in a quadratic form

$$\frac{1}{2}\dot{\Theta}^T K(\Theta)\dot{\Theta}, \tag{2.5}$$

where $K(\Theta)$ is a matrix that describes the mass distribution of a manipulator as a function of the joint vector Θ. Each element of $K(\Theta)$ must have units of inertia (kgm^2). Clearly, $K(\Theta)$ must be positive definite so that the quantity in (2.5) is always positive and represents energy. Furthermore, $K(\Theta)$ must be symmetric by the following argument: In a quadratic form such as (2.5) above, any skew symmetric matrix can be added to $K(\Theta)$ without changing the value of the quadratic form (the energy). Since any matrix can be written as the sum of a symmetric matrix plus a skew symmetric matrix, we conclude that the skew symmetric portion of $K(\Theta)$ must be zero since otherwise the system could contain mass that does not contribute to the system's kinetic energy. Hence $K(\Theta)$ is symmetric. We will call $K(\Theta)$ the **kinetic energy matrix** of the manipulator.

Also, it should be clear intuitively that it is possible to describe the potential energy of a manipulator by a scalar function of joint positions only, say $P(\Theta)$. We can then write the Lagrangian of the system as

$$L = \frac{1}{2}\dot{\Theta}^T K(\Theta)\dot{\Theta} - P(\Theta). \tag{2.6}$$

If we then derive the dynamic equation (2.1) of the manipulator by using Lagrange's method:

$$\frac{d}{dt}\left(\frac{\partial L}{\partial \dot{\Theta}}\right) - \frac{\partial L}{\partial \Theta} = T, \tag{2.7}$$

we obtain

$$\frac{d}{dt}\left(K(\Theta)\dot{\Theta}\right) - \frac{\partial}{\partial \Theta}\left(\frac{1}{2}\dot{\Theta}^T K(\Theta)\dot{\Theta} - P(\Theta)\right) = T, \tag{2.8}$$

or,

$$\dot{K}(\Theta)\dot{\Theta} + K(\Theta)\ddot{\Theta} - \frac{1}{2}\dot{\Theta}^T\left(\frac{\partial}{\partial \Theta}K(\Theta)\right)\dot{\Theta} + \frac{\partial}{\partial \Theta}P(\Theta) = T. \tag{2.9}$$

If we compare (2.1) and (2.9) we discover (by equating coefficients of $\ddot{\Theta}$) that

$$M(\Theta) = K(\Theta). \tag{2.10}$$

That is, the manipulator mass matrix is the kinetic energy matrix.

Furthermore, by equating terms in $\dot{\Theta}$ in (2.1) and (2.9) we have

$$V(\Theta, \dot{\Theta}) = \dot{M}(\Theta)\dot{\Theta} - \frac{1}{2}\dot{\Theta}^T\left(\frac{\partial}{\partial \Theta}K(\Theta)\right)\dot{\Theta}, \tag{2.11}$$

from which it can be shown (Koditschek [40]) that

$$\dot{M}(\Theta) = 2V_m(\Theta, \dot{\Theta}) - J, \tag{2.12}$$

where J is some skew symmetric matrix. This can be useful in that it implies that the following relationship between quadratic forms exists:

$$\frac{1}{2}X^T\dot{M}(\Theta)X = X^T V_m(\Theta, \dot{\Theta})X. \tag{2.13}$$

One further important property of the manipulator mass matrix is that all dependence on Θ comes in the form of the trigonometric functions sine and cosine. That is, in (2.4) dependence on Θ in the f_{ji} is always in the form of sines or cosines of the θ_i. Since sine and cosine are bounded for any value of their arguments, and they appear only in the numerators of the elements of $M(\Theta)$, $M(\Theta)$ is bounded for all Θ.

We can now state several properties of $M(\Theta)$.

(1) It is symmetric.

(2) It is positive definite and bounded above and below.

(3) Its inverse exists and is positive definite and bounded.

(4) Its time derivative is given by $2V_m(\Theta, \dot{\Theta}) - J$, where J is some skew symmetric matrix.

(5) It is also the kinetic energy matrix of the manipulator, i.e., the kinetic energy can be written $\frac{1}{2}\dot{\Theta}^T M(\Theta)\dot{\Theta}$.

Property (2) can be written

$$\alpha_m I_n \leq M(\Theta) \leq \beta_m I_n, \tag{2.14}$$

where I_n is the $n \times n$ identity matrix and the ordering is in the sense of positive definite matrices with $0 < \alpha_m < \beta_m$ scalars.

2.2.2 The Centrifugal and Coriolis Terms, $V(\Theta, \dot{\Theta})$

From (2.11) it is clear that it is possible to write the vector of centrifugal and Coriolis terms as the product

$$V(\Theta, \dot{\Theta}) = V_m(\Theta, \dot{\Theta})\dot{\Theta}. \tag{2.15}$$

It is also possible to write $V(\Theta, \dot{\Theta})$ in the form

$$V(\Theta, \dot{\Theta}) = \begin{bmatrix} \dot{\Theta}^T V_1(\Theta)\dot{\Theta} \\ \dot{\Theta}^T V_2(\Theta)\dot{\Theta} \\ \vdots \\ \dot{\Theta}^T V_n(\Theta)\dot{\Theta} \end{bmatrix}, \tag{2.16}$$

where the $V_i(\Theta)$ are $n \times n$ symmetric matrices. Clearly then,

$$V_m(\Theta, \dot{\Theta}) = \begin{bmatrix} \dot{\Theta}^T V_1(\Theta) \\ \dot{\Theta}^T V_2(\Theta) \\ \vdots \\ \dot{\Theta}^T V_n(\Theta) \end{bmatrix}. \tag{2.17}$$

Now note that $V(\Theta, \dot{\Theta})$ can also be written

$$V(\Theta, \dot{\Theta}) = V_p(\Theta) V_v(\dot{\Theta}) \dot{\Theta}, \tag{2.18}$$

where $V_v(\dot{\Theta})$ is the $\frac{1}{2} n(n+1) \times n$ matrix given by

$$V_v(\dot{\Theta}) = \begin{bmatrix} & \dot{\theta}_1 I_n \\ O_{n-1,1} & \dot{\theta}_2 I_{n-1} \\ \vdots & \vdots \\ O_{1,n-1} & \dot{\theta}_n \end{bmatrix}, \tag{2.19}$$

where $O_{n,m}$ is the $n \times m$ zero matrix, and $V_p(\Theta)$ is an $n \times \frac{1}{2} n(n+1)$ matrix.

It is sometimes useful to write the velocity dependent term, $V(\Theta, \dot{\Theta})$, in yet a different form [41]:

$$V(\Theta, \dot{\Theta}) = B(\Theta) \left[\dot{\Theta}\dot{\Theta} \right] + C(\Theta) \left[\dot{\Theta}^2 \right], \tag{2.20}$$

where $B(\Theta)$ is a matrix of dimensions $n \times n(n-1)/2$ of Coriolis coefficients and $\left[\dot{\Theta}\dot{\Theta} \right]$ is a $n(n-1)/2 \times 1$ vector of joint velocity products given by

$$\left[\dot{\Theta}\dot{\Theta} \right] = \left[\dot{\theta}_1\dot{\theta}_2 \ \dot{\theta}_1\dot{\theta}_3 \ \cdots \ \dot{\theta}_{n-1}\dot{\theta}_n \right]^T \tag{2.21}$$

and $C(\Theta)$ is a $n \times n$ matrix of centrifugal coefficients and $\left[\dot{\Theta}^2 \right]$ is an $n \times 1$ vector given by

$$\left[\dot{\theta}_1^2 \ \dot{\theta}_2^2 \ \cdots \ \dot{\theta}_n^2 \right]^T. \tag{2.22}$$

Again, dependence on Θ appears only in terms of sine and cosine functions, so that $V(\Theta, \dot{\Theta})$ has a bound that is independent of Θ, but increases quadratically with $\dot{\Theta}$.

We can now state several properties of $V(\Theta, \dot{\Theta})$.

(1) It is quadratic in $\dot{\Theta}$.

(2) It may be written in several factorizations, notably (2.15) through (2.22).

(3) It is related to the time derivative of the mass matrix by $V_m(\Theta, \dot{\Theta}) = \frac{1}{2}(\dot{M}(\Theta) + J)$, where J is some skew symmetric matrix.

2.2.3 The Friction Terms, $F(\dot{\Theta})$

Friction terms are complex and are probably represented only approximately by any deterministic model. We have indicated that friction has a functional dependence on $\dot{\Theta}$, although due to eccentricities in gearing and other effects, there may be a relatively important dependence on joint position as well.

The only two properties we will ascribe to our model of friction $F(\dot{\Theta})$ are

(1) Friction is a local effect, so $F(\dot{\Theta})$ is uncoupled in the sense that each element could be written $f_i(\dot{\theta}_i)$.

(2) The friction forces are dissipative, that is, $\dot{\Theta}^T F(\dot{\Theta}) \geq 0$. Another way to say this is that each function $f_i(\dot{\theta}_i)$ lies in the first and third quadrants only.

In some analyses we will assume that the friction forces are largely viscous, and we will write

$$F(\dot{\Theta}) = F_m \dot{\Theta} + T_{df}, \tag{2.23}$$

where F_m is a diagonal matrix of viscous friction coefficients, and T_{df} is unstructured friction effects.

2.2.4 The Gravity Terms, $G(\Theta)$

Again, dependence on Θ appears only in terms of sine and cosine functions in the numerators of its elements, so that $G(\Theta)$ has a bound that is independent of Θ.

2.3 The Control Method of Arimoto

Arimoto and Miyazaki [39] proposed that a manipulator modeled as

$$T = M(\Theta)\ddot{\Theta} + V_m(\Theta, \dot{\Theta})\dot{\Theta} + G(\Theta) \tag{2.24}$$

be controlled with the control law

$$T = G(\Theta) + K_p E - K_d \dot{\Theta}, \tag{2.25}$$

where K_p is an $n \times n$ diagonal matrix of position gains, K_d is an $n \times n$ diagonal matrix of damping gains, and $E = \Theta_d - \Theta$ is an $n \times 1$ vector of errors. Note that this controller does not force the manipulator to follow a trajectory, but moves the manipulator to a goal point along a path specified by the manipulator dynamics, and then regulates the position there. The control law requires that only the gravity model be known.

Arimoto and Miyazaki [39] and Koditschek [40] have given proofs of the global asymptotic stability of this control scheme. Consider the candidate Lyapunov function (see Appendix B for a brief review of Lyapunov stability theory):

$$v = \frac{1}{2}\dot{\Theta}^T M(\Theta)\dot{\Theta} + \frac{1}{2}E^T K_p E. \tag{2.26}$$

Differentiating yields

$$\begin{aligned}
\dot{v} &= \frac{1}{2}\dot{\Theta}^T \dot{M}(\Theta)\dot{\Theta} + \dot{\Theta}^T M(\Theta)\ddot{\Theta} - E^T K_p \dot{\Theta} \\
&= \frac{1}{2}\dot{\Theta}^T \dot{M}(\Theta)\dot{\Theta} - \dot{\Theta}^T K_d \dot{\Theta} - \dot{\Theta}^T V_m(\Theta, \dot{\Theta})\dot{\Theta} \\
&= -\dot{\Theta}^T K_d \dot{\Theta}
\end{aligned} \tag{2.27}$$

which is nonpositive. Note that we made use of the identity equation (2.13). Next, we can show asymptotic stability if

$$\dot{\Theta} = 0 \quad \forall t \quad \rightarrow \quad E = 0 \tag{2.28}$$

so that the system cannot get "stuck" with nonzero error. Since $\dot{v}$ can only remain zero along trajectories that have $\dot{\Theta} = 0$ and $\ddot{\Theta} = 0$, we see that in this case

$$M^{-1}(\Theta)K_p E = 0, \tag{2.29}$$

and since M^{-1} and K_p are nonsingular, we have that the largest invariant set is $E = 0$. Hence control law (2.25) applied to the system (2.24) achieves global asymptotic stability. The proof can be expanded to apply to the case where viscous friction is added to (2.24).

This proof is important in that it explains, to some extent, why today's industrial robots work. Most industrial robots use a simple error-driven servo, occasionally with gravity models, and so are quite similar to (2.25).

Much of this book is concerned with the *computed torque servo scheme*, which attempts to use a more complete model in the control law. The main reason for going to such a scheme with a more complete model is to be able to follow a specified trajectory through space with small errors.

2.4 Nonlinear Model-Based Control of Manipulators

Here we quickly summarize how a nonlinear dynamic model of a manipulator can be used in the control of the manipulator. For a slower development, see Chapter 8 of Craig [42].

The manipulator is modeled as a set of n rigid bodies connected in a serial chain with friction acting at the joints. The vector equation of motion of such a device is given by (2.1), i.e.,

$$T = M(\Theta)\ddot{\Theta} + V(\Theta, \dot{\Theta}) + F(\dot{\Theta}) + G(\Theta) + T_d. \qquad (2.30)$$

To control the manipulator, the following control law is used:

$$T = \hat{M}(\Theta)\ddot{\Theta}^* + \hat{V}(\Theta, \dot{\Theta}) + \hat{F}(\dot{\Theta}) + \hat{G}(\Theta), \qquad (2.31)$$

where the quantities with "hats" are computed from estimates of the true parameters, and

$$\ddot{\Theta}^* = \ddot{\Theta}_d + K_v\dot{E} + K_pE. \qquad (2.32)$$

In (2.32), the servo error E is defined as

$$E = \Theta_d - \Theta \qquad (2.33)$$

and K_v and K_p are $n \times n$ constant, diagonal gain matrices with $k_{vj} > 0$ and $k_{pj} > 0$ on the diagonals. Equation (2.31) is sometimes referred to as the *computed torque method* of manipulator control. Any desired trajectory of the free end of the manipulator is assumed known and expressible as time functions of joint positions, velocities, and accelerations, $\Theta_d(t), \dot{\Theta}_d(t)$, and $\ddot{\Theta}_d(t)$. Such a trajectory may be preplanned by several well-known schemes [42].

The control law (2.31) is chosen because in the favorable situation of perfect knowledge of parameter values, and no disturbances, the closed-loop error dynamics are given by

$$\ddot{E} + K_v\dot{E} + K_pE = M^{-1}(\Theta)T_d. \tag{2.34}$$

Hence in this ideal situation, the k_{vj} and k_{pj} may be chosen to place closed-loop poles of each joint, and disturbance rejection will be uniform over the entire workspace of the manipulator.

Chapter 3

ROBUSTNESS OF MODEL-BASED CONTROL OF MANIPULATORS

3.1 Introduction

The nonlinear model-based manipulator control system is a good approach to manipulator control when parameters are known and the control computer is sufficiently powerful. Such a control formulation yields a controller that suppresses disturbances and tracks desired trajectories uniformly in all configurations of the manipulator. However, this desirable performance is only achieved when complete knowledge of the manipulator parameters is available. However, these parameters are inevitably in error so that system performance never reaches the ideal, and in fact, such a *detuned* system may be unstable.

In this chapter we discuss the robustness of the nonlinear model-based control scheme that was presented in Chapter 2. Here we assume that parameter values are known only as a set of bounds. For example, as concerns the true value of the ith parameter p_i we know

$$p_i \in [l_i, h_i]. \tag{3.1}$$

In discussing the robustness of the nonlinear model-based controller introduced in Chapter 2, there are three major issues that we might wish to address:

(1) When the bounds are quite small (corresponding to good knowledge of parameters) we would like to show that the system performs well. That

19

is, we hope that the control scheme we presented in Chapter 2 *degrades gracefully* when knowledge of parameters is good but not perfect.

(2) We would like to show that if the bounds are quite large, then the system is still at least stable, although naturally the quality of control will degrade. That is, we are interested in determining how large the parameter bounds may be while still ensuring a stable control system.

(3) With an eye toward improving the robustness to modeling errors, we might wish to redesign the servo portion $\ddot{\Theta}^*$ of the controller.

Because of the complex nonlinear nature of the system, our analysis will result in various bounds that represent *sufficient* conditions for a certain level of performance. Hence, for the first issue to be investigated we seek a sufficient condition (satisfiable if parameter knowledge is "good") such that servo errors remain "small." For the second issue we seek a sufficient condition (satisfiable even when parameter knowledge is "mediocre") such that the control system is at least stable. Work on the third issue is intended to improve robustness by showing that with a certain servo law, and bounds on parameters, control is improved relative to the servo law given in Chapter 2.

Our main focus in this chapter will be on the first issue. That is, we will show that the control scheme we presented in Chapter 2 *degrades gracefully* when knowledge of parameters is good but not perfect. We will develop bounds on servo error as a function of parameter error. We will discuss point (2), but it remains an open problem. Work related to topic (3) by other authors will be mentioned.

3.2 Previous Related Work

Questions of robustness of the computed torque scheme arise because of parameter mismatch. When we have poor knowledge of parameters, the computed torque scheme may not decouple and linearize, but may in fact cause the system to be unstable. It is interesting to view the control scheme of Arimoto and Miyazaki [39] (see Section 2.3) as the computed torque servo with

$$\hat{M}(\Theta) = I_n,$$
$$\hat{V}(\Theta, \dot{\Theta}) = 0,$$
$$\hat{F}(\dot{\Theta}) = 0,$$
$$\hat{G}(\Theta) = G(\Theta),$$

where the desired trajectory is simply a fixed point Θ_d. Viewed as a computed torque servo, this control scheme is seen to use quite a bad estimate of the dynamic model, and yet has been shown to be globally asymptotically stable. Such a proof of stability does not yet exist for other "de-tuned" computed torque servos. Arimoto and Miyazaki's work also deals with the addition of an integral term to the control law, as well as a stability proof for a Cartesian-based controller with similar structure.

The method of robustness analysis to be used in this chapter most closely follows that of Spong and Vidyasagar [43], in that it can be viewed as an application of the small-gain theorem (see Desoer and Vidyasagar [48]). However, Spong and Vidyasagar assumed that the velocity terms quadratic in $\dot{\Theta}$ could be bounded with a linear bound. This assumption is only justified if a later check is done to ensure that over the range of $\dot{\Theta}$ in question, the linear bound holds. In the analysis here, the quadratic bound is used directly, and in fact is responsible for much of the interesting structure of the problem.

A very interesting approach aimed at redesigning the servo portion of the computed torque controller in order to improve robustness is that of Slotine [45]. A modified sliding-mode approach is used to ensure the stability of the control scheme in the presence of simplified dynamic models and unmodeled effects such as resonances. Other work in the area of robustness of nonlinear manipulator control systems is that of Ha and Gilbert [46], and Egeland [47].

3.3 Error Equation with Parameter Mismatch

Assuming only structured uncertainty, the dynamic model of the manipulator (2.1) can be parameterized in terms of an $r \times 1$ vector P of parameters. These parameters are combinations of various physical quantities associated with the manipulator, such as link inertia, centers of gravity, friction coefficients, etc. For each element p_i of P, we know bounds in the form given in (3.1).

Our model-based controller uses estimates of these parameters $\hat{P}$ with errors in these estimates denoted as

$$\tilde{P} = P - \hat{P}. \tag{3.2}$$

Hence, we have a bound on the magnitude of any parameter error

$$|\tilde{p}_i| \le h_i - l_i, \tag{3.3}$$

where $\tilde{p}_i$ is the ith element of $\tilde{P}$.

Note that all the quantities in (2.2) depend on P, so it is useful to keep in mind that (2.2) might be written

$$T = M(P, \Theta)\ddot{\Theta} + Q(P, \Theta, \dot{\Theta}) + T_d. \tag{3.4}$$

Likewise, the control law of (2.31) might be written

$$T = M(\hat{P}, \Theta)\ddot{\Theta}^* + Q(\hat{P}, \Theta, \dot{\Theta}). \tag{3.5}$$

We will generally leave out the dependence on P or $\hat{P}$ in these terms, and simply write $M(\Theta)$ for the true mass matrix and $\hat{M}(\Theta)$ for the mass matrix computed with the estimated parameters, and so forth.

When errors in the parameters exist, the error equation is no longer the ideal given in (2.34). Here we derive the system equation when parameter estimates do not match the true values. We start by equating the plant (2.2) and controller (2.31) equations when they are in the form of (3.4) and (3.5), respectively:

$$M(\Theta)\ddot{\Theta} + Q(\Theta, \dot{\Theta}) + T_d = \hat{M}(\Theta)\ddot{\Theta}^* + \hat{Q}(\Theta, \dot{\Theta}). \tag{3.6}$$

Letting $\tilde{Q} = Q - \hat{Q}$ and $\tilde{M} = M - \hat{M}$,

$$\hat{M}(\Theta)\ddot{\Theta}^* = \hat{M}(\Theta)\ddot{\Theta} + \tilde{M}(\Theta)\ddot{\Theta} + \tilde{Q}(\Theta, \dot{\Theta}) + T_d, \tag{3.7}$$

or

$$\hat{M}(\Theta)\left(\ddot{\Theta}^* - \ddot{\Theta}\right) = \tilde{M}(\Theta)\ddot{\Theta} + \tilde{Q}(\Theta, \dot{\Theta}) + T_d. \tag{3.8}$$

If $\hat{M}^{-1}(\Theta)$ exists, we have

$$\ddot{E} + K_v \dot{E} + K_p E = \hat{M}^{-1}(\Theta) \left[\tilde{M}(\Theta)\ddot{\Theta} + \tilde{Q}(\Theta, \dot{\Theta}) + T_d \right]. \qquad (3.9)$$

Along with ensuring that $\hat{M}^{-1}(\Theta)$ exists, it is usually reasonable to force $\hat{M}(\Theta)$ to be positive definite (a stronger condition), although this will not be a requirement for our first robustness result. In this form of the error equation it is easy to see that the system performs well if all parameter errors are zero, since the right-hand side becomes equal to only $M^{-1}(\Theta)T_d$. However, (3.9) hides the fact that the right-hand side is a function of E, $\dot{E}$, and $\ddot{E}$, and so when parameter errors are nonzero performance is difficult to intuit, and without more careful analysis even stability cannot be concluded. We will now manipulate (3.9) to bring out all dependence on the error and its derivatives.

For the remainder of this section and occasionally in the book, we will drop the arguments Θ and $\dot{\Theta}$ in the interest of brevity, and simply write M, V, G, etc. Note that if the arguments are other than Θ and $\dot{\Theta}$ we will show them. For example, $V(\Theta, \dot{\Theta}_d)$ will appear (the second argument is the desired velocity rather than the actual velocity).

Using

$$\hat{M}^{-1}\tilde{M}\ddot{\Theta} = \hat{M}^{-1}\tilde{M}\ddot{\Theta}_d - \hat{M}^{-1}\tilde{M}\ddot{E} \qquad (3.10)$$

and

$$I + \hat{M}^{-1}\tilde{M} = \hat{M}^{-1}M, \qquad (3.11)$$

we remove $\ddot{\Theta}$ from (3.9) to yield

$$\hat{M}^{-1}M\ddot{E} + K_v\dot{E} + K_p E = \hat{M}^{-1}\left[\tilde{M}\ddot{\Theta}_d + \tilde{Q} + T_d\right]. \qquad (3.12)$$

Now, we multiply both sides by $M^{-1}\hat{M}$ to obtain

$$\ddot{E} + M^{-1}\hat{M}K_v\dot{E} + M^{-1}\hat{M}K_p E = M^{-1}\left[\tilde{M}\ddot{\Theta}_d + \tilde{Q} + T_d\right], \qquad (3.13)$$

or

$$\ddot{E} + M^{-1}\hat{M}K_v\dot{E} + M^{-1}\hat{M}K_p E = M^{-1}\left[\tilde{M}\ddot{\Theta}_d + \tilde{V} + \tilde{F} + \tilde{G} + T_d\right]. \qquad (3.14)$$

From (2.16) we know that the ith component of $\tilde{V}(\Theta, \dot{\Theta})$ can be written as $\dot{\Theta}^T \tilde{V}_i(\Theta)\dot{\Theta}$. Since

$$\dot{\Theta} = \dot{\Theta}_d - \dot{E}, \tag{3.15}$$

the ith term of $\tilde{V}(\Theta, \dot{\Theta})$ is $(\dot{\Theta}_d - \dot{E})^T \tilde{V}_i(\Theta)(\dot{\Theta}_d - \dot{E})$. Expanding this to four terms and taking advantage of the fact that the $\tilde{V}_i(\Theta)$ are symmetric, leads to

$$\tilde{V}(\Theta, \dot{\Theta}) = \tilde{V}(\Theta, \dot{\Theta}_d) - 2\tilde{V}_m(\Theta, \dot{\Theta}_d)\dot{E} + \tilde{V}(\Theta, \dot{E}). \tag{3.16}$$

Next, using (2.23) we will write the friction model in terms of a diagonal matrix of viscous coefficients, with the other friction effects contributing to T_d:

$$\tilde{F}(\dot{\Theta}) = \tilde{F}_m \dot{\Theta}. \tag{3.17}$$

This assumption is conservative in that all friction effects are dissipative, whereas since T_d is not structured its size will reduce the stability margin in our later analysis. In other parts of this book we will not always restrict the modeled portion of the friction to be linear in nature.

We define U to be the right-hand side of (3.14), that is,

$$U = M^{-1}\left[\tilde{M}\ddot{\Theta}_d + \tilde{V} + \tilde{F} + \tilde{G} + T_d\right], \tag{3.17}$$

which, using (3.16) and (3.17), we now write as

$$U = M^{-1}\left[\tilde{M}\ddot{\Theta}_d + \tilde{V}(\Theta, \dot{\Theta}_d) + \tilde{G} + T_d + (\tilde{F}_m - 2\tilde{V}_m(\Theta, \dot{\Theta}_d))\dot{E} + \tilde{V}(\Theta, \dot{E})\right]. \tag{3.18}$$

The form of (3.18) is useful in that U shows no dependence on $\dot{\Theta}$ or $\ddot{\Theta}$, and dependence on Θ only occurs in the form of $\sin(\theta_i)$ or $\cos(\theta_i)$. Hence, U has a bound that is independent of the trajectory $(\Theta, \dot{\Theta}, \ddot{\Theta})$.

In the remainder of this chapter we study the stability of

$$\ddot{E} + M^{-1}\hat{M}K_v\dot{E} + M^{-1}\hat{M}K_p E = U,$$
$$U = M^{-1}\left[\tilde{M}\ddot{\Theta}_d + \tilde{V}(\Theta, \dot{\Theta}_d) + \tilde{G} + T_d + (\tilde{F}_m - 2\tilde{V}_m(\Theta, \dot{\Theta}))\dot{E} + \tilde{V}(\Theta, \dot{E})\right]. \tag{3.19}$$

Studying stability means (as always) studying the stability of a differential equation — for example, the differential equation given as the first part of (3.19) when driven by the input U. Note that there are alternate formulations of (3.19) in which terms from the left side of the differential equation are moved into U and vice versa. Regardless, our method of showing stability will always involve developing an input–output bound for some differential equation, and developing a bound on the driving input.

3.4 An Approach to Robustness Analysis

We will now write (3.19) in a form such that the left-hand side is a simple linear, uncoupled differential equation. This is done by writing

$$
\begin{aligned}
M^{-1}\hat{M}K_v &= K_v + M^{-1}\hat{M}K_v - K_v \\
M^{-1}\hat{M}K_p &= K_p + M^{-1}\hat{M}K_p - K_p,
\end{aligned}
\tag{3.20}
$$

which leads to

$$
\ddot{E} + K_v\dot{E} + K_pE = \xi
$$
$$
\xi = U + \left(I - M^{-1}\hat{M}\right)K_v\dot{E} + \left(I - M^{-1}\hat{M}\right)K_pE.
\tag{3.21}
$$

Note that other such expressions are possible, for example, if we used

$$
M^{-1}\hat{M}K_v = K + M^{-1}\hat{M}K_v - K,
\tag{3.22}
$$

where K was chosen to minimize $\|K - M^{-1}\hat{M}K_v\|$, we might accomplish the decoupling and linearization of the left-hand side of (3.19) while minimizing what has to be added to U. However, it can be shown that little is to be gained by this approach, and the choice of partitioning given by (3.20) leads to a result that is more simply stated.

The left-hand side of (3.21) represents a linear decoupled differential equation with input ξ as the forcing function. We will develop expressions for the L_∞ gain of the operators $\mathcal{H} : \xi \mapsto E$ and $\mathcal{G} : \xi \mapsto \dot{E}$.

For each joint, consider the L_∞ gain of the operator $\mathcal{H}_i : \xi_i \mapsto e_i$ and write the transfer function from input ξ_i to output e_i as

$$
\frac{e_i(s)}{\xi_i(s)} = h_i(s) = \frac{1}{s^2 + k_{vi}s + k_{pi}}.
\tag{3.23}
$$

Also, consider the L_∞ gain of the operator $\mathcal{G}_i : \xi_i \mapsto \dot{e}_i$ and write the transfer function from input ξ_i to output $\dot{e}_i$ as

$$\frac{\dot{e}_i(s)}{\xi_i(s)} = g_i(s) = \frac{s}{s^2 + k_{vi}s + k_{pi}}. \tag{3.24}$$

We can then bound the gain of these transfer function operators (see Appendix A) as

$$\|\mathcal{H}_i\|_\infty = \int_0^\infty |h_i(t)|dt \tag{3.25}$$

and

$$\|\mathcal{G}_i\|_2 = \int_0^\infty |g_i(t)|dt, \tag{3.26}$$

where $h_i(t)$ and $g_i(t)$ are the impulse responses of (3.23) and (3.24), respectively. Note that for simplicity we will assume that K_v and K_p have been chosen so that each joint is critically damped, i.e., $k_{vi}^2 = 4k_{pi}$. We will also assume that the gains are chosen equal on all joints. Evaluating (3.25) leads to the result

$$\|\mathcal{H}_i\|_\infty = \frac{1}{k_p}. \tag{3.27}$$

Evaluating (3.26) leads to

$$\|\mathcal{G}_i\|_\infty = \frac{4e^{-1}}{k_v}. \tag{3.28}$$

It then follows directly that bounds on the multi-input multi-output (MIMO) gain are

$$\|\mathcal{H}\|_\infty = \frac{1}{k_p} = \beta_1 \tag{3.29}$$

and

$$\|\mathcal{G}\|_\infty = \frac{4e^{-1}}{k_v} = \beta_2. \tag{3.30}$$

Hence, for zero initial conditions, we have

$$\begin{aligned} \|E\|_{T\infty} &\le \beta_1 \|\xi\|_{T\infty} \\ \|\dot{E}\|_{T\infty} &\le \beta_2 \|\xi\|_{T\infty} \end{aligned}, \tag{3.31}$$

where $\|\xi\|_{T\infty}$ denotes the L_∞^n norm of $\xi(t)$ truncated at time T. Note that from (3.31) $\xi \in L_{\infty e}^n$ implies $E, \dot{E} \in L_{\infty e}^n$, and furthermore, if $\xi \in L_\infty^n$ then $E, \dot{E} \in L_\infty^n$.

Now turning our attention to the right-hand side of (3.21) we develop a bound on $\|\xi\|_{T\infty}$ as a function of $\|E\|_{T\infty}$ and $\|\dot{E}\|_{T\infty}$. We will assume that a smooth, bounded desired trajectory is specified so that Θ_d, $\dot{\Theta}_d$, and $\ddot{\Theta}_d$ are all elements of L_∞^n. In this case, the first term on the right-hand side of (3.16) has a bound that is independent of $\dot{E}$. The second term can be bounded in the form $B\|\dot{E}\|_{T\infty}$, where $B = \| - 2V_m(\Theta, \dot{\Theta}_d)\|_{i\infty}$. The last term, which is quadratic in $\dot{E}$, has its ith element in the form $\dot{E}^T V_i(\Theta)\dot{E}$, so

$$\|\dot{E}^T V_i(\Theta)\dot{E}\|_{T\infty} \leq \|\dot{E}^T\|_{T\infty}\|V_i(\Theta)\|_{i\infty}\|\dot{E}\|_{T\infty} , \\ \leq C_i\|\dot{E}\|_{T\infty}^2 \tag{3.32}$$

where $C_i = \|V_i(\Theta)\|_{i\infty}$. Given a bound $C_i\|\dot{E}\|_{T\infty}^2$ on the ith element of an n-vector, it is clear that an L_∞ bound on the vector is

$$C = \left(\max_i C_i\right)\|\dot{E}\|_{T\infty}^2. \tag{3.33}$$

Hence, we have developed a quadratic bound on $V(\Theta, \dot{\Theta})$ in the form

$$\|V(\Theta, \dot{\Theta})\|_{T\infty} = A + B\|\dot{E}\|_{T\infty} + C\|\dot{E}\|_{T\infty}^2. \tag{3.34}$$

It is now clear that we can bound ξ as follows:

$$\|\xi\|_{T\infty} \leq \alpha_1 + \alpha_2\|E\|_{T\infty} + \alpha_3\|\dot{E}\|_{T\infty} + \alpha_4\|\dot{E}\|_{T\infty}^2, \tag{3.35}$$

where

$$\begin{aligned} \alpha_1 &= \|M^{-1}\left[\tilde{M}\ddot{\Theta}_d + \tilde{V}(\Theta, \dot{\Theta}_d) + \tilde{G} + T_d\right]\|_\infty, \\ \alpha_2 &= \|(I_n - M^{-1}\hat{M})K_p\|_{i\infty}, \\ \alpha_3 &= \|(I_n - M^{-1}\hat{M})K_v + M^{-1}(\tilde{F}_m - 2\tilde{V}_m(\Theta, \dot{\Theta}_d))\|_{i\infty}, \\ \alpha_4 &= \|M^{-1}\|_{i\infty}\left(\max_i C_i\right). \end{aligned} \tag{3.36}$$

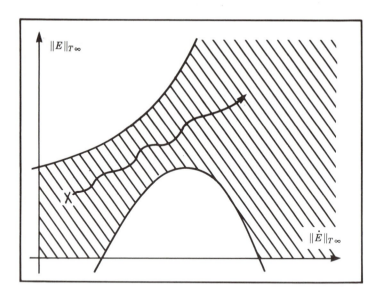

Figure 3.1 *Quadratic Half-Plane Constraints*

Hence, $E \in L_{\infty e}^n$ and $\dot{E} \in L_{\infty e}^n$ implies $\xi \in L_{\infty e}^n$. We now wish to find a condition under which $E \in L_\infty^n$ and $\dot{E} \in L_\infty^n$, which implies (by 3.35) that $\xi \in L_\infty^n$ and hence the system in (3.19) is L_∞ stable.

Combining (3.31) and (3.35) leads to two quadratic inequalities, each of which specifies a half-plane (in the $(\|E\|_{T\infty}, \|\dot{E}\|_{T\infty})$ plane) within which the truncated error magnitudes must lie:

$$-\frac{\alpha_4}{\alpha_2}\|\dot{E}\|_{T\infty}^2 + \left(\frac{1 - \beta_2\alpha_3}{\beta_2\alpha_2}\right)\|\dot{E}\|_{T\infty} - \frac{\alpha_1}{\alpha_2} \leq \|E\|_{T\infty} \qquad (3.37)$$

and

$$\left(\frac{\beta_1\alpha_4}{1 - \beta_1\alpha_2}\right)\|\dot{E}\|_{T\infty}^2 + \left(\frac{\beta_1\alpha_3}{1 - \beta_1\alpha_2}\right)\|\dot{E}\|_{T\infty} + \left(\frac{\beta_1\alpha_1}{1 - \beta_1\alpha_2}\right) \geq \|E\|_{T\infty}, \quad (3.38)$$

where we have assumed that

$$\beta_1\alpha_2 < 1,$$
$$\beta_2\alpha_3 < 1. \qquad (3.39)$$

Based on knowledge of the signs of the coefficients of the two quadratic half-plane constraints (3.37 and 3.38), we show them qualitatively in Figure 3.1. Each parabola represents an equation when we use the equal sign. We know that $\|E\|_{T\infty}$ and $\|\dot{E}\|_{T\infty}$ specify a point in the indicated region outside of both parabolas. Note that because we are dealing with norms, only the first quadrant of the plane is of interest. As shown in Figure 3.1, there is no closed region in which errors are bounded. For example, Figure 3.1 shows the possible evolution of servo error magnitude from the origin. Clearly, error magnitude might grow to infinity (or might not). However, under certain circumstances, the two parabolas will intersect and create a closed region bordering the origin as shown in Figure 3.2.

To determine the condition under which a closed region in the magnitude plane exists (as shown in Figure 3.2), we equate the two quadratic equations that describe the two half-plane boundaries. This leads to another quadratic, which must have two real roots if the parabolas are to intersect as shown in Figure 3.2. This condition (that the discriminant of this quadratic equation be positive) can be written as

$$\beta_1\alpha_2 + \beta_2\alpha_3 + 2\beta_2\sqrt{\alpha_1\alpha_4} < 1. \tag{3.40}$$

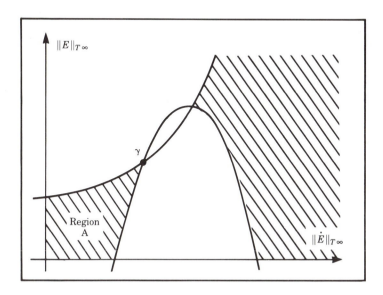

Figure 3.2 *Quadratic Half-Plane Constraints*

Note that this inequality contains the earlier assumptions of (3.39). Therefore if (3.40) is satisfied, then $E \in L_\infty^n$ and $\dot{E} \in L_\infty^n$ because $\|E\|_{T\infty} \le e_{max}$ and $\|\dot{E}\|_{T\infty} \le \dot{e}_{max}$, where $(e_{max}, \dot{e}_{max})$ are the coordinates of the point γ in Figure 3.2. Note that $\|E\|_{T\infty}$ and $\|\dot{E}\|_{T\infty}$ are both *continuous* functions of time because they are each the magnitude of the output of an integrator whose input is a continuous function. Therefore it is not possible to "jump" from one shaded region to the other in Figure 3.2. Hence, when (3.40) is satisfied, the system in (3.19) is L_∞ stable.

3.5 Statement of Robustness Theorem 3.1

We are now ready to state a robustness theorem.

Theorem 3.1

For a manipulator described by (2.1) under control of the nonlinear model-based control law of (2.31) with errors in parameter values:

If

(1) $\hat{M}^{-1}(\Theta)$ exists,

(2) $k_{vi}^2 = 2k_{pi} > 0$,

(3) Θ_d and its first two derivatives are bounded,

(4) The disturbances T_d are uncorrelated, bounded,

(5) (3.40) holds,

(6) $\|E\| = \|\dot{E}\| = 0$ at $t = t_0$,

Then

The system is L_∞ stable with $\|E\|_\infty$ and $\|\dot{E}\|_\infty$ remaining within region A in Figure 3.2.

Examining (3.40) we see that a requirement for this condition to be met is that

$$\|I_n - M^{-1}\hat{M}\| < \frac{1}{2} - \epsilon, \tag{3.41}$$

where ϵ is quite small and goes to zero as k_v increases. This turns out to be quite restrictive, and will only be met if the parameter errors are relatively small. However, if it is met, then Theorem 3.1 provides an excellent bound (given by region A) on the size of the error.

Hence, Theorem 3.1 gives us the result that for "good" knowledge of parameters (but not perfect), the nonlinear model-based controller performs very well.

3.6 Numerical Example of Theorem 3.1

A simple two-degree-of-freedom manipulator (as shown in Figure 3.3) was considered in a numerical example. The manipulator was modeled as two rigid links (of lengths l_1, l_2) with point masses at the distal ends of the links (m_1, m_2). It moves in a vertical plane with gravity acting. Such a manipulator, although quite simple, is subject to joint torques due to inertial, centrifugal, Coriolis, gravity, and frictional effects.

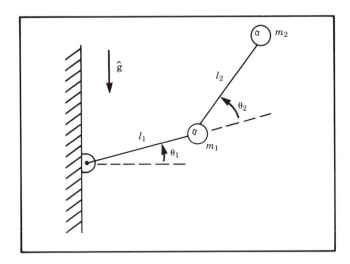

Figure 3.3 *Manipulator Considered for Robustness Calculation*

The equations of motion for this device are (Craig [42]):

$$
\begin{aligned}
\tau_1 = {}& m_2 l_2^2 (\ddot{\theta}_1 + \ddot{\theta}_2) + m_2 l_1 l_2 c_2 (2\ddot{\theta}_1 + \ddot{\theta}_2) + (m_1 + m_2) l_1^2 \ddot{\theta}_1 \\
& - m_2 l_1 l_2 s_2 \dot{\theta}_2^2 - 2 m_2 l_1 l_2 s_2 \dot{\theta}_1 \dot{\theta}_2 + m_2 l_2 g s_{12} \\
& + (m_1 + m_2) l_1 g s_1 + v_1 \dot{\theta}_1 + k_1 sgn(\dot{\theta}_1),
\end{aligned}
\tag{3.42}
$$

$$
\begin{aligned}
\tau_2 = {}& m_2 l_1 l_2 c_2 \ddot{\theta}_1 + m_2 l_1 l_2 s_2 \dot{\theta}_1^2 + m_2 l_2 g s_{12} + m_2 l_2^2 (\ddot{\theta}_1 + \ddot{\theta}_2) \\
& + v_2 \dot{\theta}_2 + k_2 sgn(\dot{\theta}_2),
\end{aligned}
$$

where $c_1 = \cos(\theta_1)$, $s_{12} = \sin(\theta_1 + \theta_2)$, etc. The parameters v_i and k_i are viscous and Coulomb friction coefficients, respectively.

The following values were used.

$$
\begin{aligned}
\|\dot{\Theta}_d\|_\infty &= 2.0 \quad \text{rad/sec} \\
\|\ddot{\Theta}_d\|_\infty &= 6.0 \quad \text{rad/sec}^2 \\
k_p &= 225.0 \\
k_v &= 30.0 \\
m_1 &\in [5.7, 6.3] \quad \text{Kg} \\
m_2 &\in [2.85, 3.15] \quad \text{Kg} \\
l_1 &\in [0.999, 1.001] \quad \text{M} \\
l_2 &\in [0.999, 1.001] \quad \text{M}
\end{aligned}
\tag{3.43}
$$

Because the presence of friction can only increase stability (by dissipating energy), the friction coefficients were left out of this numerical example. However, the effect of misknowledge in their values could easily have been included here as well. That is, error in mass parameters is $\pm 5\%$, and error in link lengths is $\pm 0.1\%$. With these values the robustness condition of Theorem 3.1 was met (the left-hand side of (3.40) evaluated to 0.744). Figure 3.4 shows the regions of interest in the error magnitude plane. In the worst case we have

$$
\begin{aligned}
\|E\|_\infty &\leq 0.024 \quad \text{rad}, \\
\|\dot{E}\|_\infty &\leq 0.269 \quad \text{rad/sec}.
\end{aligned}
\tag{3.44}
$$

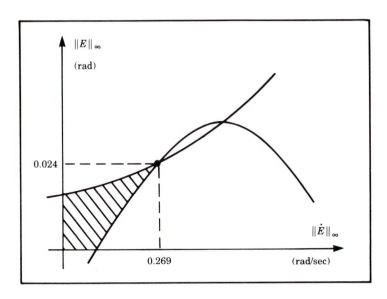

Figure 3.4 *Regions for Numerical Example of Section 3.6*

That is, for all trajectories with desired velocity and acceleration bounded as given in (3.43), worst-case position error is about 1.3 degrees, and worst-case velocity error is about 14 degrees per second.

The numerical computation performed requires the computation of all the α_i and the β_i so that (3.40) can be evaluated. Furthermore, the points where the parabolas (Figure 3.2) cross the axes, as well as where the parabolas intersect are computed. The input data are the range of uncertainty in parameter values, the servo gains, and the bounds on the velocity and acceleration of the desired trajectory. In computing the α_i it is helpful to note that all extremum will occur when joint angles are a multiple of 90 degrees, and when the parameters are at an extrema of their ranges.

3.7 A Robustness Conjecture

Robustness Theorem 3.1 is quite conservative. That is, for moderately mis-tuned parameters, it will fail to conclude stability when in fact, the system may be stable. In this section we present a conjecture concerning the robustness of the computed torque servo.

Based on many simulations and also on trials with actual manipulators, the computed torque servo actually appears to be extremely robust to parameter errors. It is quite common to see reasonable response in simulation when inertial and other parameters are in error by 50% or even 100%. Discussions with other researchers indicates similar results in their experiments. However, there has been no published proof of such robustness.

After a significant amount of time spent on the problem, we were unable to discover a proof less conservative than that of Theorem 3.1. However, all empirical evidence supports the following:

Conjecture 3.1

If

(1) $\hat{M}(\Theta) > 0$ and symmetric,

(2) $K_v > \alpha^* I_n$,

Then

      The system is L_∞ stable.

That is, if the estimate of the manipulator mass matrix is positive definite and symmetric,† and if the velocity feedback is sufficiently large and positive, the system will be stable despite possibly large parameter errors. The intention here is that α^* is not terrifically large — that is, the conjecture is not intended to propose "nearly infinite gain" in order to ensure stability. The proof of this conjecture, as well as an expression or a bound on α^*, awaits further research.

The conjecture is believed to be true because of the following argument. The system under study is (3.19), which has the form

$$\ddot{E} + A(t)\dot{E} + B(t)E = U(t). \qquad (3.45)$$

Various related scalar systems (sometimes called Lienard's equation) have been proved to be Lagrange stable (i.e., all solutions bounded) under relatively simple conditions on $a(t)$ and $b(t)$ (as a minimum, $a(t) > \alpha^* > 0$

---

† A slightly weaker conjecture would require that the product $M^{-1}\hat{M}$ be positive definite.

and $b(t) > 0$ is generally required), for example, see LaSalle and Lefschetz [50]. In our case, we must consider a vector differential equation (3.45) in which $A(t)$ and $B(t)$ have all positive eigenvalues. Depending on the details of how (3.45) is set up, $U(t)$ may or may not depend on $E(t)$. In the work done leading to Theorem 3.1, such dependence in $U(t)$ led eventually to two parabolas, which were required to intersect to show stability. It appears that under other formulations one parabola degenerates into a half-plane, and so stability might be shown with conditions as simple as those stated in the conjecture.

It should be noted that the system being addressed is the rigid-link case with a continuous-time servo, etc. — all the assumptions stated earlier in this book. If Conjecture 3.1 is proved, it will mean that the computed torque servo is a good control scheme for rigid-link manipulators with a sufficiently high servo rate. That is, if there is a better way to control a manipulator, it must have to do with flexibilities and other effects, rather than any intrinsic fault with the nonlinear model-based control method.

Chapter 4

REVIEW OF
ADAPTIVE CONTROL

4.1 Introduction

Traditionally, control systems have been designed based on a good understanding of the system to be controlled. When knowledge of the system is limited the relatively modern issues of robust control, adaptive control, and learning control become important.

One way to attempt to deal with poor knowledge of parameters in a control scheme is through techniques that are generally called *adaptive control*. Adaptive control is closely related to the problem of system identification; in fact, generally an adaptive controller can be viewed as being composed of two parts:

(1) An identification portion, which identifies parameters of the plant itself, or parameters that appear in the controller for the plant.

(2) A control law portion, which implements a control law that is in some way a function of the parameters identified.

Hence, adaptive control can be viewed as an identification scheme coupled with a control scheme. The central problem in the synthesis of adaptive controllers is to prove rigorously that the resulting overall system is stable.

Adaptive-control strategies take on many forms. For the case of a linear plant and unknown constant parameters, methodologies have become established for designing adaptive-control systems. Among these methodologies, the two most widely described are the *self tuning regulator* and the *model reference adaptive control* scheme. Both schemes are quite similar in many ways. We will briefly describe model reference adaptive control in general as an introduction to reviewing the previous work in adaptive control of manipulators in particular.

4.2 Model Reference Adaptive Control

In this section we briefly review the theory of model reference adaptive control (MRAC) as applied to linear plants. We restrict our attention to plants that are continuous-time, single-input–single-output (SISO), minimum-phase, and of relative degree one.† Our emphasis here is on describing the structure of the controller and showing the form of a rigorous proof of stability for the overall scheme. For a more complete exposition, see Sastry [63] and references therein.

In model reference adaptive control, the desired closed-loop performance of the plant is described by specifying a *reference model* that exhibits the desired dynamic response. The adaptive controller observes output error between the desired and actual plants and, based on this error, updates a vector of parameters used by the controller in order to reduce the error. Let the unknown plant transfer function be in the form

$$P(s) = k_p \frac{n_p(s)}{d_p(s)}, \tag{4.1}$$

where $n_p(s)$ and $d_p(s)$ are monic, coprime polynomials of degree $n - 1$ and n, respectively. Furthermore, the zeros of n_p are assumed to be in the left half-plane, and k_p is assumed to be positive. The input and output of the plant are denoted by $u(t)$ and $y_p(t)$, respectively. The reference model that specifies the eventual desired behavior of the system is

$$M(s) = k_m \frac{n_m(s)}{d_m(s)}, \tag{4.2}$$

where $n_m(s)$ and $d_m(s)$ are monic, coprime polynomials of degree $n - 1$ and n, respectively, n_m has no right half-plane zeros, and k_m is positive. The

---

† *Relative degree* refers to the difference between the number of poles and zeros in a transfer function.

input and output of the reference model are denoted by $r(t)$ and $y_m(t)$, respectively. The error between plant and reference model is

$$e_1 = y_p(t) - y_m(t). \tag{4.3}$$

The structure of the model reference controller is shown in Figure 4.1. Note that if $e_1 = 0$ then the plant is tracking the reference input $r(t)$ exactly as the reference model $M(s)$ does. The plant transfer function is altered by $2n$ tunable parameters c_0, C, d_0, D, which appear in what might be called a "precompensator" block F_1 and a "feedback controller" block F_2 (C and D are $n - 1 \times 1$ vectors). Each of these blocks contains an auxiliary signal generator, in F_1

$$v^{(1)} = (sI - \Lambda)^{-1}b\, u, \tag{4.4}$$

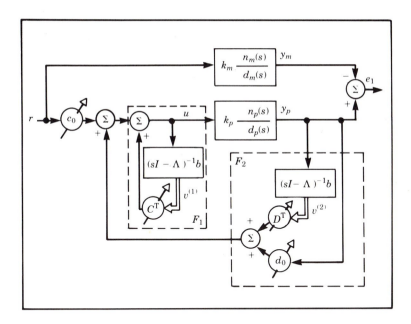

Figure 4.1 *Structure of an MRAC Controller (from Sastry [63])*

and in F_2

$$v^{(2)} = (sI - \Lambda)^{-1}b\, y_p, \tag{4.5}$$

where $v^{(1)}$ and $v^{(2)}$ are $n \times 1$ vectors of signals.

Given the structure of the controller shown in Figure 4.1, it is not difficult to show (see Sastry [63]) that there exist values for c_0, C, d_0, D, which we will call c_0^*, C^*, d_0^*, D^*, such that the transfer function of the plant plus controller equals $M(s)$, the reference model. Furthermore, this can be done with $\Lambda(s) = det(sI - \Lambda)$, a stable polynomial. Hence, there exists a setting for the tunable parameters such that reference-model tracking is achieved and all internal signals of the controller are bounded if the input $r(t)$ is bounded. Defining

$$C^*(s) = c_1^* + c_2^* s + \cdots + c_{n-1}^* s^{n-2} \tag{4.6}$$

and

$$D^*(s) = d_0^* \Lambda(s) + d_1^* + d_2^* s + \cdots + d_{n-1}^* s^{n-2}, \tag{4.7}$$

then this perfect tuning is given by

$$c_0^* = \frac{k_m}{k_p},$$
$$C^*(s) = n_m(s) - n_p(s), \tag{4.8}$$
$$D^*(s) = \frac{d_p(s) - d_m(s)}{k_p}.$$

If the plant transfer function was known, (4.8) could be used to achieve perfect model matching. Design of the adaptive law begins with forming a state–space description of the controller of Figure 4.1. The section of the controller that corresponds to the plant plus its controller can be represented in state–space by a realization of the form

$$k_p \frac{n_p(s)}{d_p(s)} = c_p^T (sI - A_p)^{-1} b_p, \tag{4.9}$$

with the state vector $\begin{bmatrix} x_p^T & v^{(1)T} & v^{(2)T} \end{bmatrix}^T$, where x_p is the $n \times 1$ state vector of the unknown plant. Likewise, it is possible to write the reference model in state–space form using a nonminimal representation of the same form, namely

$$k_m \frac{n_m(s)}{d_m(s)} = c_m^T (sI - A_m)^{-1} b_m, \tag{4.10}$$

with the state vector $\left[x_m^T \ v_m^{(1)T} \ v_m^{(2)T}\right]^T$, where x_m is the $n \times 1$ state vector of the reference model. Finally, subracting these two state–space representations yields the *error equation*, which relates error in tunable parameters to errors in model-tracking error. The error equation can be put in the form

$$\dot{e} = A_m e + b_m \phi^T w,$$
$$e_1 = c_m e, \tag{4.11}$$

where ϕ is the $2n \times 1$ vector of errors in the tunable parameters given by

$$\phi^T = \left[c_0, C^T, d_0, D^T\right] - \left[c_0^*, C^{*T}, d_0^*, D^{*T}\right] \tag{4.12}$$

and w is the $2n \times 1$ vector of signals given by

$$w^T = \left[r, v^{(1)T}, y_p, v^{(2)T}\right]. \tag{4.13}$$

Note that (4.11) describes a SISO linear system driven by the input $\phi^T w$. We now develop a means of adjusting our estimates of these parameters so that model-tracking error $e_1(t)$ goes to zero asymptotically. Derivation of the adaptive law and the associated stability proof require that the transfer function of the reference model be strictly positive real (SPR). A brief review of strictly positive real transfer functions is given in Appendix C. Note that in the relative-degree-one case, it is always possible to obtain an SPR reference model, although in some cases a prefiltering technique is required (Sastry [63]). Because $c_m^T(sI - A_m)b_m$ is assumed SPR, there exist positive definite matrices P and Q such that

$$A_m^T P + P A_m = -Q,$$
$$P b_m = c_m^T. \tag{4.14}$$

Now condsider the Lyapunov candidate function

$$v(e, \phi) = e^T P e + \phi^T \phi. \tag{4.15}$$

Upon differentiating, if we choose

$$\dot{\phi} = -e_1 w, \tag{4.16}$$

we are led to

$$\dot{v}(E, \phi) = -e^T Q e, \qquad (4.17)$$

which is nonpositive. Hence, e and ϕ are bounded and

$$\int_0^\infty -\dot{v}\,dt < \infty. \qquad (4.18)$$

From (4.11) it is easily seen that if $r(t)$ is bounded then $\dot{e}$ is bounded, and so e is uniformly continuous, and hence, $\dot{v}(e, \phi)$ is uniformly continuous. Because $\dot{v}(e, \phi)$ always has the same sign, is uniformly continuous, and has a finite integral, we conclude [53] that

$$\lim_{t \to \infty} \dot{v}(e, \phi) = 0, \qquad (4.19)$$

and so

$$\lim_{t \to \infty} e = 0. \qquad (4.20)$$

Note that

$$\lim_{t \to \infty} \dot{\phi} = 0 \qquad (4.21)$$

and

$$\lim_{t \to \infty} v(e, \phi) = \lim_{t \to \infty} \phi^T \phi = v^*, \qquad (4.22)$$

but we cannot yet conclude that ϕ converges to zero. To ensure the convergence of ϕ to zero, we must consider the question of sufficient excitation of the system. This problem has been studied by Morgan and Narendra [66], Anderson [67], and Kreisselmeier [68], who have shown that for a given $w(t)$, the system in (4.11) is a linear time-varying system that is uniformly asymptotically stable if and only if there exists an α and a δ both positive, such that for all t_0

$$\int_{t_0}^{t_0+\delta} ww^T dt \geq \alpha I_{2n}, \qquad (4.23)$$

where I_{2n} is the $2n \times 2n$ identity matrix. The condition in (4.23) is referred to as a *sufficient richness* condition. It essentially requires that $w(t)$ move around so as to span the $2n$ dimensional space over any interval of length δ. An interesting result (Boyd and Sastry [57]) is that if the spectrum of the reference signal $r(t)$ has $2n$ spectral lines, then the sufficient richness condition (4.23) will be met.

Our analysis thus far has assumed that a set of parameter values exist so that the plant plus controller will match the reference model in transfer function — that is, that all uncertainty in our model is *structured uncertainty*. In practice, it will always be impossible to obtain a perfect parametric model of the plant, and so our assumption is suspect. Recently, research in adaptive control has largely been focused on this problem of unstructured uncertainty, or *unmodeled dynamics* and external disturbances. It was shown (Rohrs [58], Ioannou [59], Peterson and Narendra [60]) that when some unstructured uncertainty was present, a loss of stability or unbounded parameter estimates could result.

One method for ensuring robustness of the MRAC adaptive scheme to external disturbances is to add a dead zone to the update law. The intuitive idea is to turn off adaptation if errors are sufficiently small that they might have arisen from unstructured uncertainty rather than from parameter mistuning. The size of the dead zone is related to the size of the upper bound on the uncertainty. As this dead zone increases in size the benefits of parameter-adaptive control are diminished. MRAC adaptive control therefore is suited to situations in which a fairly good parametric model of the plant is available.

A second method of making MRAC schemes robust has to do with ensuring a certain level of persistent excitation (Narendra and Annaswamy [56]). It can be shown that if the system is sufficiently excited, a bounded disturbance cannot cause unbounded solutions to the equations describing the system. In this scheme, an appropriate measure of the excitation of the system must exceed some measure of the upper bound of the disturbance.

In conclusion, a theory for MRAC of linear plants with constant coefficients has been developed over the last several years. The theory applies rigorously only to systems with parameters that are unknown constants. In practice, such MRAC systems may perform well with systems in which the parameters are slowly varying. It should be noted that so-called *hyperstability theory* is sometimes used to prove such schemes stable. It is important to note that application of hyperstability theory does not lead to anything significantly different from the Lyapunov stability result obtained in this section. It is simply a different (yet related) method of showing stability. Finally, with MRAC and every other form of adaptive system, some notion of sufficient excitation must arise. Clearly, the system must be made to "wiggle around" in a sufficiently complete way if the effects of all system parameters are to be observed and those parameters estimated.

4.3 Review of Previous Work in Adaptive Manipulator Control

Note that standard MRAC theory, as reviewed in the last section, is not applicable to plants other than linear time-invariant systems, and hence is unsuitable for direct application to the manipulator control problem. The same comments apply to self-tuning regulator (STR) theory. Despite this observation, virtually all parameter-adaptive schemes for adaptive control of manipulators to date have relied on the standard MRAC or STR structure. Stability proofs have been the standard ones (similar to the method reviewed in the last section) along with time-scale arguments claiming "slow variations" in the parameters to be identified. However, modern manipulators move sufficiently fast that, for example, the effective joint inertia at a given joint may change by 300% in a fraction of a second. Hence, most previously published schemes for adaptive control of mechanical manipulators are on somewhat weak theoretical footing. It is also the case that previous work on adaptive control of manipulators almost completely ignores the issue of persistent excitation.

Previous work seems to have been motivated by the desire to avoid computation in the control computer. That is, in lieu of computing the complex dynamic model, it was desired to use some adaptive scheme to track changes in an assumed linear plant. This is not a reasonable objective, because

(1) Resulting designs have no theoretical stability proof because the underlying theory is valid only for constant unknown parameters.

(2) Computing the dynamic model of a manipulator as a part of the control law should no longer be regarded as prohibitive.

Hence, rather than to avoid computation, a nobler goal for adaptive control systems is to improve on performance relative to nonadaptive systems.

Dubowsky and DesForges [83] were perhaps the first to propose the application of MRAC to the manipulator-control problem. Their adaptation scheme is based on linear decoupled models, one per joint. They state that the underlying theory is valid only if the manipulator changes configuration slowly relative to the adaptation rate. Horowitz and Tomizuka [90, 91, 135, 136] have proposed a scheme that takes some of the manipulator dynamics into account. They write the dynamics with portions depending only on manipulator position treated as unknown parameters that are adaptively identified. Hence, the rate at which the manipulator changes configuration must be low compared to the adaptation time constants in order for the theory to be valid.

Recent extensions to the scheme seem to have improved on this situation [92]. Anex and Hubbard [69] implemented a variant of the scheme of Horowitz and Tomizuka. Koivo et al. [96-103] have used discrete linear time-invariant decoupled models, one per joint, to model the dynamics of the manipulator. Based on this assumption, an adaptive controller based on STR theory has been proposed. There is no rigorous proof of stability because the parameters are not constants, and the assumption of "slowly varying parameters" is questionable. Leininger [113-115] and Backes [70-72] and their colleagues likewise assume linear decoupled models for each joint and then proceed to apply STR ideas — least-squares identification with a forgetting factor, and a pole-placement controller. Again, no stability proof exists for these schemes due to the nonlinearities in the manipulator dynamics.

Stoten [131] does not handle nonlinearities in the manipulator dynamics and so applies an adaptation law without proof of stability. Durrant-Whyte [86] has to make the assumption that the rate of evolution of configuration is small compared to the absolute parameter values in order to make some approximations, hence the stability proof is not rigorous. Liu [119] shows another scheme that is based completely on the assumption of linear models with slowly varying parameters, hence there is no stability proof for the manipulator problem. Kubo and Ohmae [105] implement an adaptive controller in a Cartesian space-control scheme, but make gross approximations by neglecting terms in the dynamics. Elliot et al. [87] have not been rigorous in forming a valid discrete-time model of a two-link manipulator, and then use an update scheme intended for linear systems. Lee et al. [107-111] linearize the dynamics about the desired trajectory in order to obtain a linear system. However, the resulting linear system is of course a time-varying one, and so again, the underlying theory is only valid for "slow" manipulator motions. LeBorgne et al. [106] base their work on linear theory, and admit that it is valid only to the extent that the "slowly varying" assumption is true. Nicolo and Katende [125] apply the linear MRAC formulation to a manipulator and so do not have a rigorous stability proof. Kim and Shin

[94] admit in a footnote that their scheme is valid only in the case of "slow variations," which means the manipulator must be limited in speed. Walters [137] bases his work on SISO linear models for each joint, and hence has no stability proof for the actual situation. Lim and Eslami [116-118] must make an approximation that a certain time-varying matrix is constant, and so their parameter-estimation–style scheme is on somewhat weak footing. Neuman and Stone [124] discuss the variations in parameters of a linearized model of a manipulator, but do not suggest a theory by which to handle parameter identification in such a case. The paper of Choi et al. [79] extends the work of Lee and Chung, but it is still based on a linearization that leads to a time varying-system for which the adaptive theory is not rigorous. Seraji [127] bases his system on a linear model, and so a rigorous proof of stability is not possible. Ma and Lee [120] improve slightly on the work of Lee et al., but do not repair the underlying "slow-variation" assumption. Goor [88] has an interesting approach, but does not appear to handle the nonlinearities present in the manipulator dynamics rigorously. Koditschek [95] suggests an interesting problem, but does not give a full solution. Hsia [93], in a review paper, discusses only the work of others.

Another group of work, which sometimes goes under the name of adaptive control of manipulators, is based on sliding-mode or variable-structure systems. In these schemes, parameters are not identified, and so these may or may not be called adaptive-control schemes. Balestrino et al. [73-75] use what they call the "unit vector" adaptation scheme, which produces a switching signal that is injected into the control torque to achieve good tracking despite poor knowledge of parameters. Such chattering controls (which in theory switch at infinite frequency) cannot be implemented, and the derivative of servo error goes to zero only in the mean. Such high-frequency control action may also excite unmodeled resonances. Finally, as mentioned earlier, such schemes do not provide any identification of parameters.

Nicosia and Tomei [126] have similarly suggested a sliding-mode–style "adaptive" scheme for manipulators. In a brief paper, Singh [130] describes a switching-signal-synthesis adaptive scheme based on the work of Corless and Leitmann [61] for uncertain systems. Singh's work is restricted to manipulators with revolute joints. The signal-synthesis scheme of Lim and Eslami [116-118] is also in the class of high-frequency switching-signal schemes. Finally, there is some interesting work on modified sliding-mode control of manipulators (Slotine [45]). Slotine and Coetsee [128, 129] have recently described a very interesting scheme that combines some true parameter adaptation along with sliding-mode control action.

4.4 Conclusion

To date, the majority of work on the problem of adaptively controlling a mechanical manipulator has simply been the application to the manipulator system of methods that were developed for linear systems. Such methods do not lead to an overall adaptive scheme that can be rigorously proven stable. It has been the aim of many researchers to use adaptation as a means to simplify or otherwise avoid doing the nonlinear model computations in the control computer. As computing power becomes more available and schemes to compute the dynamic model are improved, this motivation diminishes. A more interesting use of adaptive control is to attempt to outperform the simpler nonadaptive controllers.

Chapter 5

ADAPTIVE CONTROL
OF MANIPULATORS

5.1 Introduction

The "computed torque" servo was introduced in Chapter 2 as a method of using the dynamic model of a manipulator in a control-law formulation. Such
a control formulation yields a controller that suppresses disturbances and
tracks desired trajectories uniformly in all configurations of the manipulator.
However, this desirable performance is contingent on two assumptions that
have made implementations of the computed torque servo less than ideal.
First, the dynamic model of the manipulator must be computed quickly
enough so that discretization effects do not degrade performance relative to
the continuous-time, zero-delay ideal. Second, the values of parameters appearing in the dynamic model in the control law must match the parameters
of the actual system if the beneficial decoupling and linearizing effects of the
computed torque servo are to be realized.

Some recent work in formulating efficient computational algorithms for manipulator dynamics, along with the increase in the performance/price ratio
of computing hardware, have caused the first difficulty of employing the computed torque servo to diminish [20, 23, 27]. The work reported in this chapter
is intended to address the second difficulty, that of imprecise knowledge of
manipulator parameters.

We have seen that the issue of robustness of the computed torque servo is
somewhat an open question. However, we will see that when an adaptive

element is added to the scheme, the algorithm is globally asymptotically stable subject to some easily stated criteria.

In this chapter we present an adaptive scheme of manipulator control that takes full advantage of any known parameters while estimating the remaining unknown parameters. The overall adaptive control system maintains the structure of the computed torque servo, but in addition has an adaptive element. After sufficient on-line learning, the control algorithm decouples and linearizes the manipulator so that each joint behaves as an independent second-order system with fixed dynamics.

5.2 The Dynamic Model of a Manipulator

The manipulator is modeled as a set of n rigid bodies connected in a serial chain with friction acting at the joints. The vector equation of motion of such a device can be written in the compact form given by (2.1) as

$$T = M(\Theta)\ddot{\Theta} + Q(\Theta, \dot{\Theta}). \tag{5.1}$$

The jth element of (5.1) can be written in a sum-of-products form, just as we did in (2.4),

$$\tau_j = \sum_{i=1}^{a_j} m_{ji} f_{ji}(\Theta, \ddot{\Theta}) + \sum_{i=1}^{b_j} q_{ji} g_{ji}(\Theta, \dot{\Theta}), \tag{5.2}$$

where the m_{ji} and q_{ji} are parameters formed by products of such quantities as link masses, link inertia tensor elements, lengths, friction coefficients, and the gravitational acceleration constant. The $f_{ji}(\Theta, \ddot{\Theta})$ and the $g_{ji}(\Theta, \dot{\Theta})$ are functions that embody the dynamic structure of the manipulator's motion geometry. In this book we assume that the *structure* of these parameters and functions is known, but the numerical values of some or all of the parameters m_{ji} and q_{ji} are unknown. We will, however, assume that bounds on the parameter values are known, although these bounds may sometimes be quite loose.† This is equivalent to the situation of knowing the kinematic structure

---

† In fact, such bounds are needed only for the parameters that appear in the manipulator's mass matrix. However, for generality, we will assume that bounds on all parameters are known.

of a manipulator and having parametric models of joint friction effects, but knowing only some, or perhaps none, of the dynamic parameters such as mass distribution of the links and friction coefficients.

5.3 Nonlinear Model-Based Control

To control the manipulator, we propose the control law

$$T = \hat{M}(\Theta)\ddot{\Theta}^* + \hat{Q}(\Theta, \dot{\Theta}), \tag{5.3}$$

where $\hat{M}(\Theta)$ and $\hat{Q}(\Theta, \dot{\Theta})$ are estimates of $M(\Theta)$ and $Q(\Theta, \dot{\Theta})$ and

$$\ddot{\Theta}^* = \ddot{\Theta}_d + K_v\dot{E} + K_pE. \tag{5.4}$$

In (5.4), the servo error $E = [e_1 e_2 \ldots e_n]^T$ is defined as

$$E = \Theta_d - \Theta, \tag{5.5}$$

and K_v and K_p are $n \times n$ constant, diagonal-gain matrices with k_{vj} and k_{pj} on the diagonals. As introduced in Chapter 2, (5.3) is sometimes referred to as the *computed torque method* of manipulator control. The desired trajectory of the manipulator is assumed known as time functions of joint positions, velocities, and accelerations, $\Theta_d(t), \dot{\Theta}_d(t)$, and $\ddot{\Theta}_d(t)$.

The jth element of (5.3) can be written in the sum-of-products form

$$\tau_j = \sum_{i=1}^{a_j} \hat{m}_{ji} f_{ji}(\Theta, \ddot{\Theta}^*) + \sum_{i=1}^{b_j} \hat{q}_{ji} g_{ji}(\Theta, \dot{\Theta}), \tag{5.6}$$

where the $\hat{m}_{ji}$ and $\hat{q}_{ji}$ are estimates of the parameters appearing in (5.2).

The control law (5.3) is chosen because in the favorable situation of perfect knowledge of parameter values and no disturbances, the jth joint has closed-loop dynamics given by the error equation

$$\ddot{e}_j + k_{vj}\dot{e}_j + k_{pj}e_j = 0. \tag{5.7}$$

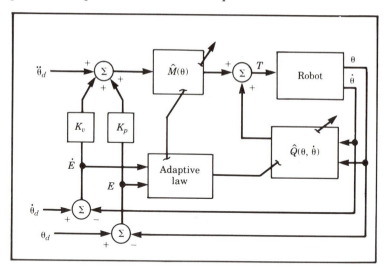

Figure 5.1 *The Controller with an Adaptive Element*

Hence in this ideal situation, the k_{vj} and k_{pj} may be chosen to place the closed-loop poles associated with each joint, and disturbance rejection will be uniform over the entire workspace of the manipulator.

Figure 5.1 is a block diagram indicating the structure of the controller that makes use of a dynamic model of the manipulator. An adaptive element is also indicated. This adaptive element observes servo errors and adjusts the parameters that appear in the control law (5.3). The remainder of this chapter is concerned with the design of this adaptive element, proof of global stability of the design, and other related issues.

We first consider what we will call the "ideal case," in which we have a perfect structural model of the manipulator dynamics. In this case, parameter errors are the sole source of nonperfect decoupling and linearization of the plant. That is, there exists a tuning (or setting) of the parameters that would cause the model in the computer to match the dynamics of the actual mechanical manipulator exactly. In a later section of this chapter we will address the problem of robustness to disturbances, in which case there is no set of parameters that will cause the model and the actual mechanism dynamics to match exactly. Obviously, some (at least small) amount of disturbance will always be present and we need to ensure that the overall adaptive system is robust to its presence.

5.4 The Error Equation

When estimates of parameters do not match the true parameter values, the closed-loop system will not perform as indicated by (5.7). By equating (5.1) and (5.3) we obtain

$$\ddot{E} + K_v\dot{E} + K_p E = \hat{M}^{-1}(\Theta) \left[\tilde{M}(\Theta)\ddot{\Theta} + \tilde{Q}(\Theta,\dot{\Theta}) \right],\qquad (5.8)$$

where $\tilde{M}(\Theta) = M(\Theta) - \hat{M}(\Theta)$ and $\tilde{Q}(\Theta,\dot{\Theta}) = Q(\Theta,\dot{\Theta}) - \hat{Q}(\Theta,\dot{\Theta})$ represent errors in the dynamic model used in the controller arising from errors in the parameters of the model.

In a given application, we may know some of the parameters m_{ji} and q_{ji}. Of the a_j parameters m_{ji} and b_j parameters q_{ji} appearing in the dynamic equation (5.2) of the jth joint, let r_j and s_j, respectively, of them be *unknown*, with $r_j \leq a_j$ and $s_j \leq b_j$ for all j. Re-index the unknown parameters (if necessary) and note that the jth component of the expression in the brackets in (5.8) can be written

$$\tilde{\tau}_j = \sum_{i=1}^{r_j} \tilde{m}_{ji} f_{ji}(\Theta,\ddot{\Theta}) + \sum_{i=1}^{s_j} \tilde{q}_{ji} g_{ji}(\Theta,\dot{\Theta}),\qquad (5.9)$$

where

$$\tilde{m}_{ji} = m_{ji} - \hat{m}_{ji},$$
$$\text{and} \qquad\qquad (5.10)$$
$$\tilde{q}_{ji} = q_{ji} - \hat{q}_{ji}$$

are parameter errors.

The error equation (5.8) relates errors in the parameter estimates to servo errors. The discussion preceding (5.9) tells how to partition the dynamics arbitrarily into known and unknown portions. This partitioning will allow us to construct an adaptive scheme that makes full use of known parameters, and adjusts only the estimates of the unknown parameters. For example, we may know the inertial properties of the manipulator, but not the friction coefficients, or we may know the parameters of some links but not others, etc.

We will write the error equation (5.8) in the form

$$\ddot{E} + K_v\dot{E} + K_p E = \hat{M}^{-1}(\Theta)\, W(\Theta,\dot{\Theta},\ddot{\Theta})\, \Phi,\qquad (5.11)$$

where Φ is an $r \times 1$ vector containing the parameter errors for all the parameters in the system, and $W(\Theta, \dot{\Theta}, \ddot{\Theta})$ is an $n \times r$ matrix of functions. For brevity, the arguments of $\hat{M}^{-1}$ and W will be dropped in the sequel. The number of system parameters is

$$r \leq \sum_{j=1}^{n}(r_j + s_j).$$ (5.12)

These r system parameters, which are the m_{ji} and the q_{ji} either alone or in combination, will now be called $P = [p_1 \; p_2 \; \cdots \; p_r]^T$ and their estimates are $\hat{P} = [\hat{p}_1 \; \hat{p}_2 \; \cdots \; \hat{p}_r]^T$, so that

$$\Phi = P - \hat{P}.$$ (5.13)

For uniformity, W and P can be defined so that each element of P is positive.

For the jth joint an error equation may be written as

$$\ddot{e}_j + k_{vj}\dot{e}_j + k_{pj}e_j = (\hat{M}^{-1}W\Phi)_j,$$ (5.14)

where $(\cdot)_j$ means the jth element of the $n \times 1$ vector, $\hat{M}^{-1}W\Phi$. Thus, in general, a parameter error for *any* parameter in the system will give rise to errors on the jth joint.

In the following analysis it will be important that the product $\hat{M}^{-1}W$ remain bounded at all times. Since W is composed of bounded functions of manipulator trajectory, W will remain bounded if the trajectory of the manipulator remains bounded. The matrix $\hat{M}(\Theta)$ will remain positive definite and invertible if we ensure that all parameters m_{ji} remain within a sufficiently small range near the actual parameter value. With this as motivation, we will restrict our estimates of the parameters to lie within bounds, such that

$$l_i - \delta < \hat{p}_i < h_i + \delta,$$ (5.15)

where we know that the actual value p_i lies between l_i and h_i, and where δ is positive and chosen such that $\hat{M}^{-1}$ remains bounded as long as (5.15) holds.

5.5 The Adaptation Algorithm

The adaptive law will compute how to change parameter estimates as a function of a filtered servo error signal. The filtered servo error for the jth joint is

$$e_{1j}(s) = (s + \psi_j)e_j(s), \tag{5.16}$$

where the ψ_j are positive constants. Hence,

$$E_1 = \dot{E} + \Psi E, \tag{5.17}$$

where $\Psi = diag(\psi_1 \ \psi_2 \ \dots \ \psi_n)$. Note that for manipulators instrumented with position and velocity sensors, the value E_1 can be computed simply from sensor readings and the filter need not be implemented as such.

The ψ_j are chosen such that the transfer function

$$\frac{s + \psi_j}{s^2 + k_{vj}s + k_{pj}} \tag{5.18}$$

is strictly positive real (SPR).† Then, by the positive real lemma [64] we are assured of the existence of the positive definite matrices P_j and Q_j such that

$$\begin{aligned} A_j^T P_j + P_j A_j &= -Q_j \\ P_j B_j &= C_j^T, \end{aligned} \tag{5.19}$$

where the matrices A_j, B_j, and C_j are the matrices of a minimal state–space realization of the filtered error equation of the jth joint

$$\begin{aligned} \dot{x}_j &= A_j x_j + B_j (\hat{M}^{-1}W\Phi)_j \\ e_{1j} &= C_j x_j, \end{aligned} \tag{5.20}$$

where the state vector is $x_j = [e_j \ \dot{e}_j]^T$.

The filtered error equation of the entire system in state–space form is given by

$$\begin{aligned} \dot{X} &= AX + B\hat{M}^{-1}W\Phi \\ E_1 &= CX, \end{aligned} \tag{5.21}$$

---

† A rational SPR function $T(s)$ is one that is analytic in the closed right half-plane and has $Re(T(j\omega)) > 0 \ \forall \omega$.

where A, B, and C are all block diagonal (with A_j, B_j, and C_j on the diagonals, respectively) and $X = [x_1 \ x_2 \ \ldots \ x_n]^T$. Forming the $2n \times 2n$ matrices $P = diag(P_1 \ P_2 \ \ldots \ P_n)$ and $Q = diag(Q_1 \ Q_2 \ \ldots \ Q_n)$, we have that $P > 0$, $Q > 0$, and

$$A^T P + P A = -Q$$
$$P B = C^T. \tag{5.22}$$

We now use Lyapunov theory to derive an adaptation law [65]. The Lyapunov function candidate

$$v(X, \Phi) = X^T P X + \Phi^T \Gamma^{-1} \Phi \tag{5.23}$$

with $\Gamma = diag(\gamma_1 \ \gamma_2 \ \ldots \ \gamma_r)$ and $\gamma_i > 0$ is nonnegative in both servo and parameter errors. Differentiation with respect to time leads to

$$\dot{v}(X, \Phi) = -X^T Q X + 2\Phi^T \left(W^T \hat{M}^{-1} E_1 + \Gamma^{-1} \dot{\Phi} \right). \tag{5.24}$$

If we choose

$$\dot{\Phi} = -\Gamma W^T \hat{M}^{-1} E_1, \tag{5.25}$$

we have

$$\dot{v}(X, \Phi) = -X^T Q X, \tag{5.26}$$

which is nonpositive because Q is positive definite. Since $\Phi = P - \hat{P}$, we have $\dot{\Phi} = -\dot{\hat{P}}$, and from (5.25) we have the adaptation law

$$\dot{\hat{P}} = \Gamma W^T \hat{M}^{-1} E_1. \tag{5.27}$$

Equations (5.23) and (5.26) imply that X and Φ are bounded. The basic update law is given by (5.27). However, in order to restrict the parameter estimates to lie within the bounds given in (5.15), we augment the update law for parameter p_i with the reset conditions

$$\begin{cases} \hat{p}_i(t^+) = l_i, & \text{if } \hat{p}_i(t) \leq l_i - \delta; \\ \hat{p}_i(t^+) = h_i, & \text{if } \hat{p}_i(t) \geq h_i + \delta. \end{cases} \tag{5.28}$$

Thus if an estimate moves outside its known bound by an amount δ, it is reset to its bound. This parameter *resetting* causes a step change in Φ in (5.21). This cannot cause an instantaneous change in X and so we can write the value of the Lyapunov function before and after the reset of p_i to its lower bound at time t_j as

$$v(t_j) = X^T P X + \sum_{\substack{k=1 \\ k \neq i}}^{r} \frac{1}{\gamma_k} \phi_k^2 + \frac{1}{\gamma_i}(p_i - l_i + \delta)^2$$

$$v(t_j^+) = X^T P X + \sum_{\substack{k=1 \\ k \neq i}}^{r} \frac{1}{\gamma_k} \phi_k^2 + \frac{1}{\gamma_i}(p_i - l_i)^2. \tag{5.29}$$

Therefore the change in v due to the resetting of $\hat{p}_i$ at time t_j is

$$-\epsilon_j = v(t_j^+) - v(t_j) = -(2(p_i - l_i) - \delta)(\frac{\delta}{\gamma_i}), \tag{5.30}$$

where ϵ_j is positive and lower bounded by $\frac{\delta^2}{\gamma_i}$. Similarly, if we reset p_i to its upper bound at time t_j, we have

$$-\epsilon_j = v(t_j^+) - v(t_j) = (2(p_i - h_i) - \delta)(\frac{\delta}{\gamma_i}), \tag{5.31}$$

where ϵ_j is positive and lower bounded by $\frac{\delta^2}{\gamma_i}$. With this addition of parameter resetting, (5.26) becomes

$$\dot{v}(X, \Phi) = -X^T Q X - \sum_{j}^{q} \delta(t - t_j)\epsilon_j, \tag{5.32}$$

where q resets take place, and $\delta(\cdot)$ here refers to the unit impulse function. Hence, the addition of parameter resetting maintains the nonpositiveness of $\dot{v}(X, \Phi)$ and hence the system is stable in the sense of Lyapunov with X and Φ bounded.

Since $X, \Phi, \hat{M}^{-1}$, and W are bounded, we see from (5.21) that $\dot{X}$ is bounded as well. Thus X is uniformly continuous, and so is $\dot{v}(X, \Phi)$. From (5.23) and (5.26) we have that

$$\lim_{t \to \infty} v(X, \Phi) = v^* \tag{5.33}$$

exists, with

$$v^* - v(X_0, \Phi_0) = -\int_0^\infty X^T Q X dt - \sum_{j=1}^{q} \epsilon_j, \tag{5.34}$$

where q parameter resettings take place. Since the left-hand side is known to be finite, and both terms on the right-hand side have the same sign, we know that each term on the right-hand side must be finite. Hence, at most a finite number q of parameter resets take place.

We know [53] that since $X^T Q X$ is positive, uniformly continuous, and has a finite integral that

$$\lim_{t \to \infty} X^T Q X = 0, \tag{5.35}$$

and thus

$$\lim_{t \to \infty} E = 0$$
$$\lim_{t \to \infty} \dot{E} = 0. \tag{5.36}$$

Hence, the adaptive scheme is stable (in the sense that all signals remain bounded) and trajectory tracking errors E and $\dot{E}$ converge to zero. As concerns convergence of the parameter errors, note that if the trajectory is not persistently exciting we can say only

$$\lim_{t \to \infty} \left| \Gamma^{-\frac{1}{2}} \Phi \right| = \sqrt{v^*}. \tag{5.37}$$

Note that $\ddot{\Theta}$, the actual acceleration of the manipulator, appears in the adaptation law of any parameter representing an inertia. Manipulators do not usually have acceleration sensors. However, the integrating action of the parameter update law reduces the necessity for good acceleration information. This has been verified in simulation and in experiments with an actual manipulator (see Sections 5.8 and 5.9).

5.6 Parameter Error Convergence

After a finite amount of time, all parameter resets have occurred, and we may write the equations describing the complete system (i.e., 5.21 and 5.25) as

$$\begin{bmatrix} \dot{X} \\ \dot{\Phi} \end{bmatrix} = \begin{bmatrix} A & BU^T \\ -\Gamma U C & 0 \end{bmatrix} \begin{bmatrix} X \\ \Phi \end{bmatrix}, \tag{5.38}$$

where $U = (\hat{M}^{-1} W)^T$. Several researchers have studied the asymptotic stability of (5.38). In [66-68] it is shown that (5.38) is uniformly asymptotically stable if the linear system (A, B, C) meets the earlier SPR condition and if U satisfies the persistent-excitation condition

$$\alpha' I_r \leq \int_{t_0}^{t_0+\rho} UU^T \, dt \leq \beta' I_r \tag{5.39}$$

for all t_0, where α', β', and ρ are all positive. Condition (5.39) says that the integral of UU^T must be positive definite and bounded over all intervals of length ρ. Note that a matrix of the form UU^T has dimension $r \times r$ but can have a rank of no greater than n (and usually, $r > n$). Hence, (5.39) means that U must vary sufficiently over the interval ρ so that the entire r-dimensional space is spanned. Note that by restricting the ranges of our estimates we have ensured that $\hat{M}$ remains invertible, and hence U is bounded, so that the right-hand inequality in (5.39) is already met.

Next we claim that because $\hat{M}$ is a bounded positive definite symmetric matrix, the left-hand inequality of (5.39) will be satisfied if for some $\alpha > 0$,

$$\alpha I_r \leq \int_{t_0}^{t_0+\rho} W^T W \, dt \tag{5.40}$$

is satisfied. A proof by contradiction of this assertion is as follows. Assume that (5.40) does not imply (5.39). Then we can always find a vector v such that for any $\gamma > 0$

$$\gamma > v^T \left[\int_{t_0}^{t_0+\rho} (\hat{M}^{-1}W)^T (\hat{M}^{-1}W) dt \right] v. \tag{5.41}$$

In particular, (5.41) holds for $\gamma = \alpha \lambda_{\hat{m}min}^2$, where

$$\lambda_{\hat{m}min} = \min_t \left[\min_i \left[\lambda_i(\hat{M}^{-1}(t)) \right] \right] > 0. \tag{5.42}$$

So,

$$\begin{aligned}
\alpha \lambda_{\hat{m}min}^2 &> v^T \left[\int_{t_0}^{t_0+\rho} (\hat{M}^{-1}W)^T (\hat{M}^{-1}W) dt \right] v \\
&= \int_{t_0}^{t_0+\rho} \|\hat{M}^{-1}Wv\|^2 dt \\
&> \lambda_{\hat{m}min}^2 \int_{t_0}^{t_0+\rho} \|Wv\|^2 dt,
\end{aligned} \tag{5.43}$$

or

$$\alpha > \int_{t_0}^{t_0+\rho} \|Wv\|^2 dt. \tag{5.44}$$

But this contradicts (5.40), and hence it is true that (5.40) implies (5.39).

Finally, since we have shown (independent of persistent excitation) that the servo error converges to zero under this control scheme, the persistent-excitation condition of (5.40) will be met if the *desired trajectory* satisfies

$$\alpha I_r \leq \int_{t_0}^{t_0+\rho} W_d^T W_d dt \leq \beta I_r, \tag{5.45}$$

where W_d is the W function evaluated along the desired rather than the actual trajectory of the manipulator. Hence we have derived a condition on the desired trajectory such that all parameters will be identified after a sufficient learning interval.

5.7 Robustness to Bounded Disturbances

When there is a disturbance present, it is likely that no tuning of the parameters will result in perfect model matching. Recently research in adaptive control has largely been focused on this problem of robustness. It was shown for the linear plant case (Rohrs [58], Ioannou [59], Peterson and Narendra [60]) that when some unstructured uncertainty was present, a loss of stability or unbounded parameter estimates could result.

For the adaptive scheme presented in this chapter, we will show that the presence of bounded disturbances will not result in loss of stability or unbounded estimates. This property is due to the fact that we have already assumed a priori bounds on the parameter values and have implemented "resetting" rules as part of the adaptive algorithm. However, of course there are adverse effects of disturbances on the adaptive scheme, as shown below:

(1) Servo errors do not converge asymptotically to zero, but rather, converge to a bounded region near zero. The size of this region is given in a straightforward manner by the choice of servo gains, and by the upper bound on the magnitude of the disturbances.

(2) Parameter estimates may not converge, but in the worst case, our resetting laws will maintain the estimates in their a priori bounds. If a

certain persistent-excitation condition is met, then the parameter errors will converge to a bounded region near zero. Unfortunately, it is difficult to state this persistent-excitation condition in terms of the desired trajectory.

To analyze the effect of disturbances, we return to (5.8). We now assume that the modeling error has two components: one due to parameter mismatch Φ and one due to disturbances $\nu'(t)$. That is,

$$\tilde{M}(\Theta)\ddot{\Theta} + \tilde{Q}(\Theta, \dot{\Theta}) = W(\Theta, \dot{\Theta}, \ddot{\Theta})\Phi + \nu', \tag{5.46}$$

where the $n \times 1$ vector function $\nu'(t)$ is completely unknown, but is upper bounded by

$$\|\nu'(t)\| < \nu'_{max}. \tag{5.47}$$

The scope of this analysis lies in what kinds of signals may be in ν'. First, ν' can certainly contain external disturbances (assumed to be uncorrelated with the state of the control system) that are bounded, but also, ν' may contain those functions of state that can be known a priori to be bounded, for example, terms such as $p_i \cos(\theta_j)$ or $p_i sgn(\dot{\theta}_j)$. Note that if ν' was known to be linear (for example) in one or more state variables, then it would not generally be possible to know a priori that it is bounded. This is the problem of robustness to unmodeled dynamics, for which a general solution is not yet known. Given (5.46), our system (from 5.21) becomes

$$\dot{X} = AX + B\hat{M}^{-1}W\Phi + \nu$$
$$E_1 = CX, \tag{5.48}$$

where $\nu = B\hat{M}^{-1}\nu'$ is $2n \times 1$ with upper bound

$$\|\nu(t)\| < \nu_{max}. \tag{5.49}$$

Note that ν_{max} is computable from ν'_{max}. Choosing the same Lyapunov candidate function as before,

$$v(X, \Phi) = X^T P X + \Phi^T \Gamma^{-1} \Phi, \tag{5.50}$$

and using the same adaptation law, we obtain

$$\dot{v}(X, \Phi) = -X^T Q X + 2X^T P \nu. \tag{5.51}$$

The previous analysis for the effect of parameter resetting (5.28 through 5.32) is still valid, and so will not be repeated here. Equation (5.51) shows that for sufficiently large X, $\dot{v}(X, \Phi)$ is nonpositive, and so X is bounded. In particular, the region in which $v(X, \Phi)$ may increase is given by

$$X < 2 \mathcal{Q}^{-1} \mathcal{P} \nu, \tag{5.52}$$

which is upper bounded by the hypersphere

$$\|X\|_2 < 2 \frac{\lambda_{pmax}}{\lambda_{qmin}} \nu_{max}, \tag{5.53}$$

where λ_{pmax} is the maximum eigenvalue of $\mathcal{P}$ and λ_{qmin} is the minimum eigenvalue of $\mathcal{Q}$. Equation (5.53) gives an upper bound on the eventual size of servo errors in the presence of bounded disturbances. Note that λ_{pmax} and λ_{qmin} are a function of the chosen servo gains, and the ψ_j from (5.18). Note that (5.53) implies that there exist constants e_{max} and $\dot{e}_{max}$ such that

$$\begin{aligned}
\|E\| &< e_{max} \\
\|\dot{E}\| &< \dot{e}_{max}
\end{aligned} \tag{5.54}$$

and from (5.8), with (5.46) substituted for the bracketed term, that $\ddot{e}_{max}$ exists such that

$$\|\ddot{E}\| < \ddot{e}_{max}. \tag{5.55}$$

Note that the above analysis does not guarantee that Φ goes to zero, or even that it remains small. In fact, without saying anything about excitation of the trajectory, Φ may diverge. However, because we have implemented resetting rules to ensure fixed upper and lower bounds for the elements of Φ, we are assured that our estimates will remain bounded at all times.

To consider persistent excitation in the presence of disturbances, we see that (5.38) becomes

$$\begin{bmatrix} \dot{X} \\ \dot{\Phi} \end{bmatrix} = \begin{bmatrix} A & BU^T \\ -\Gamma UC & 0 \end{bmatrix} \begin{bmatrix} X \\ \Phi \end{bmatrix} + \begin{bmatrix} \nu \\ 0 \end{bmatrix} \tag{5.56}$$

and note (from our previous result) that if

$$\alpha I \leq \int_{t_0}^{t_0 + \rho} W^T W \, dt, \tag{5.57}$$

then (5.56) repesents an asymptotically stable system driven with a bounded input. Hence, the state vector $[X^T \Phi^T]^T$ converges to a region that can be upper bounded by a ball centered at the origin of this $2n + r$ dimensional space. Hence, ϕ_{max} exists such that

$$\|\Phi\| < \phi_{max}. \tag{5.58}$$

At this point, we would like to give a condition for the *desired* trajectory such that (5.57) is true. Since servo errors no longer converge asymptotically to zero, this is no longer trivial. For the case of a linear plant (i.e., W linear) it has been shown [56] that if a certain measure of the excitation of the desired trajectory exceeds a measure of the disturbance, the actual trajectory will also be sufficiently exciting. For the nonlinear case of interest here, it is not yet apparent how to state this requirement simply.

We have shown that the adaptive control system designed for the ideal case of only structured uncertainty will exhibit robustness to bounded external disturbances, and to a certain class of a priori bounded unmodeled dynamics. Servo errors will be upper bounded by a value computable from the chosen servo gains, and parameter errors will remain small if a certain persistent-excitation condition is met, and in the worst case will remain within their a priori known bounds.

5.8 Simulation Results

A simple two-degrees-of-freedom manipulator (as shown in Figure 3.3) was simulated to test the adaptive algorithm. The manipulator was modeled as two rigid links (of lengths l_1, l_2) with point masses at the distal ends of the links (m_1, m_2). It moves in a vertical plane with gravity acting. Both viscous (v_i coefficients) and Coulomb friction (k_i coefficients) are simulated at the joints. This is the same manipulator as described in Section 3.6. The equations of motion for this device are given in (3.42).

These equations are in the sum-of-products form of (5.2), with

$$\begin{aligned}
a_1 &= 3, \\
b_1 &= 6, \\
a_2 &= 2, \\
b_2 &= 4.
\end{aligned} \tag{5.59}$$

We assumed that the values of the link lengths l_1 and l_2 and the value of the gravitational constant g are known. Even these parameters could be unknown, but in most realistic situations they are known quite well. Since the l_i and g do not appear as independent parameters, we have $r_j = a_j$ and $s_j = b_j$ in (5.9). Writing the system's error equation in the form given in (5.11) results in a total of six system parameters ($r = 6$ in (5.12)). The parameters are

$$
\begin{aligned}
p_1 &= m_1 \\
p_2 &= m_2 \\
p_3 &= k_1 \\
p_4 &= v_1 \\
p_5 &= k_2 \\
p_6 &= v_2
\end{aligned}
\tag{5.60}
$$

and the W matrix is

$$
W = \begin{bmatrix} w_{11} & w_{12} & w_{13} & w_{14} & 0 & 0 \\ 0 & w_{22} & 0 & 0 & w_{25} & w_{26} \end{bmatrix},
\tag{5.61}
$$

where

$$
\begin{aligned}
w_{11} &= l_1^2 \ddot{\theta}_1 + l_1 g s_1 \\
w_{12} &= (l_1^2 + l_2^2 + 2l_1 l_2 c_2)\ddot{\theta}_1 + (l_2^2 + l_1 l_2 c_2)\ddot{\theta}_2 \\
&\quad + l_1 g s_1 + l_2 g s_{12} - l_1 l_2 s_2 \dot{\theta}_2^2 - 2l_1 l_2 s_2 \dot{\theta}_1 \dot{\theta}_2 \\
w_{13} &= sgn(\dot{\theta}_1) \\
w_{14} &= \dot{\theta}_1 \\
w_{22} &= (l_2^2 + l_1 l_2 c_2)\ddot{\theta}_1 + l_2^2 \ddot{\theta}_2 + l_1 l_2 s_2 \dot{\theta}_1^2 + l_2 g s_{12} \\
w_{25} &= sgn(\dot{\theta}_2) \\
w_{26} &= \dot{\theta}_2.
\end{aligned}
\tag{5.62}
$$

The parameter update law is then as given in (5.27) and (5.28).

Parameters used in the simulation were chosen to be realistic, and process noise was added in the form of random perturbations at the torque output of the actuators. The desired trajectory had the form

$$
\begin{aligned}
\theta_{1d} &= a_1 + b_1(\sin(t) + \sin(2t)) \\
\theta_{2d} &= a_2 + b_2(\cos(4t) + \cos(6t)).
\end{aligned}
\tag{5.63}
$$

Figure 5.2 shows some results from a typical "identification paradigm" simulation. In this case, all parameters initially had substantial errors, which were corrected by the adpative controller over the first several seconds of operation. Figure 5.2a shows $\hat{m}_1$ starting from an initial value of 3 Kg and adapting to the true value of 4 Kg. Likewise, $\hat{m}_2$ changes from an initial guess of 1.5 Kg to a true value of 2 Kg. Friction parameters are also shown, for example, the estimate of the Coulomb friction coefficient of joint 2 ($\hat{k}_2$) for which the initial guess was 0.0 and the true value was 1 NtM. Also shown in Figures 5.2f and 5.2h is the servo error $E = [e_1, e_2]^T$ diminishing as the system tunes itself. The velocity error $\dot{E}$ was equally well behaved.

Figure 5.3 show some results from a typical "adaptation paradigm" simulation. In this case, all parameters were initially tuned to their true values, but the value of mass at the end of the second link m_2 undergoes a step change from 2 Kg to 3 Kg at $t = 5$ seconds. These plots indicate how the system can track a step parameter change. Note how all parameters, as well as the servo errors, are temporarily disturbed by the change.

Figure 5.4 is the same "identification paradigm" situation as in Figure 5.2, but a substantial amount of noise has been added to the torque at each joint. This is to test the robustness to an external disturbance acting at the joints. In this case the system appears quite robust to such disturbances. The magnitude of the servo error perturbations is consistent with the value of the noise amplitude divided by the closed-loop position gain.

Figure 5.5 is the same "adaptation paradigm" as in Figure 5.3, but with noise added as in the test of Figure 5.4.

The results displayed in Figure 5.6 are from an experiment designed to observe the effect of the lack of persistent excitation on the system. In this case, the manipulator is hanging straight down (so gravity causes no sag) and the desired trajectory is a constant. With the manipulator not moving, the input is not sufficiently exciting, and so the parameters tend to diverge, because of noise present in the system. Note that errors remain quite small. In this case, the parameter bounds were not implemented; normally, these bounds would prevent the parameter estimates from becoming unbounded.

Several tests were run to investigate the effect of various unmodeled effects in the manipulator dynamics, and drift in parameters. For example, Figure 5.7 shows the system initially tuned, but the viscous friction coefficient of joint 1 (v_1) is exponentially decreasing (simulating the joint's lubrication warming up). Using the same values of adaptation gains (the γ_i) as in the other tests, the system was too slow to track this change. Figure 5.7e shows $\hat{v}_1$ only slowly starting to descend toward the true value v_1 (plotted on the same axes). Figure 5.8 shows the results of the same test with γ_4 increased by a factor of 20. In Figure 5.8e we see the ability of the system to track a slowly varying parameter.

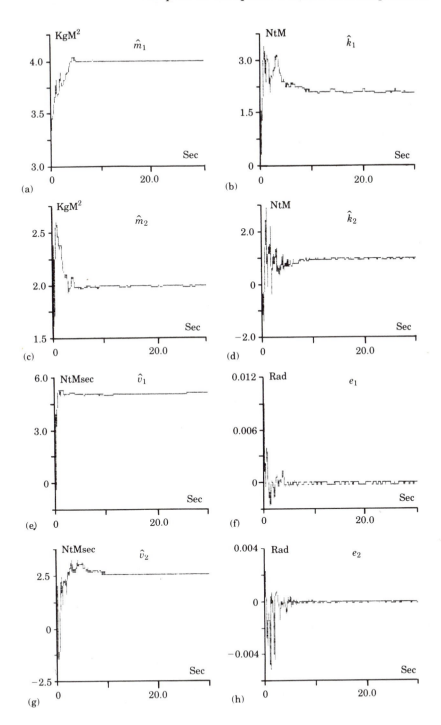

Figure 5.2 *Identification Example*

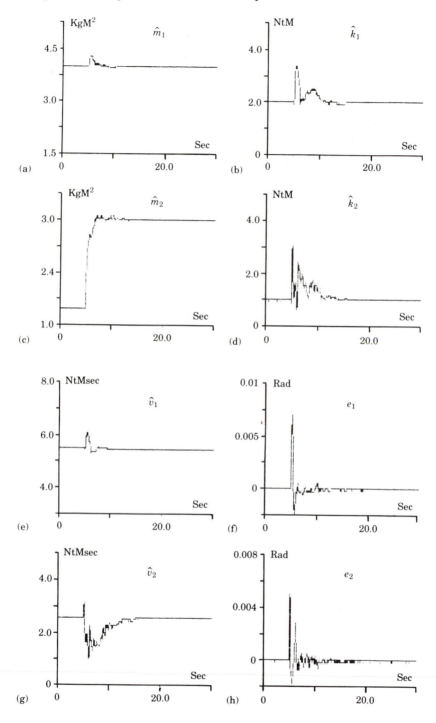

Figure 5.3 *Adaptation Example*

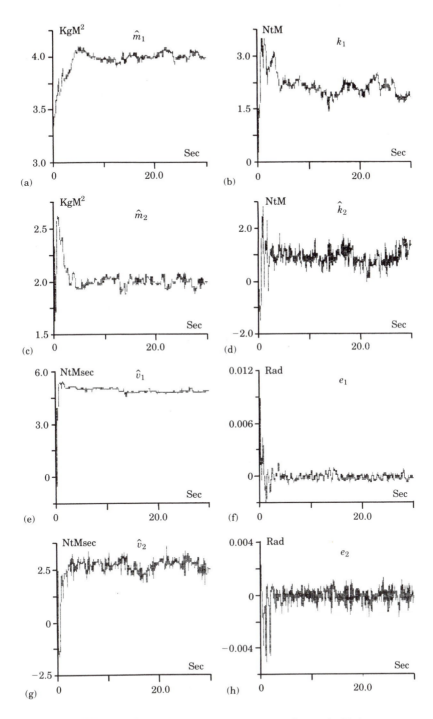

Figure 5.4 *Identification Example with Noise*

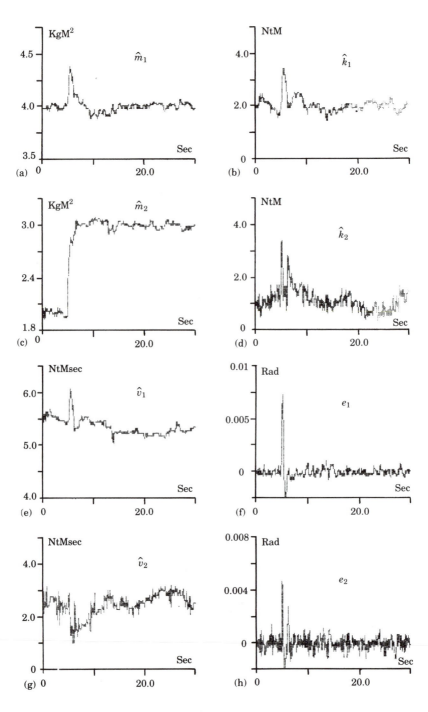

Figure 5.5 *Adaptation Example with Noise*

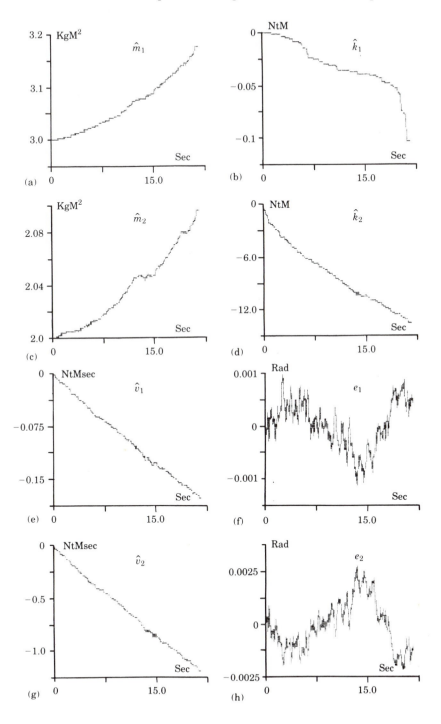

Figure 5.6 *Parameter Drift When Excitation Is Lacking*

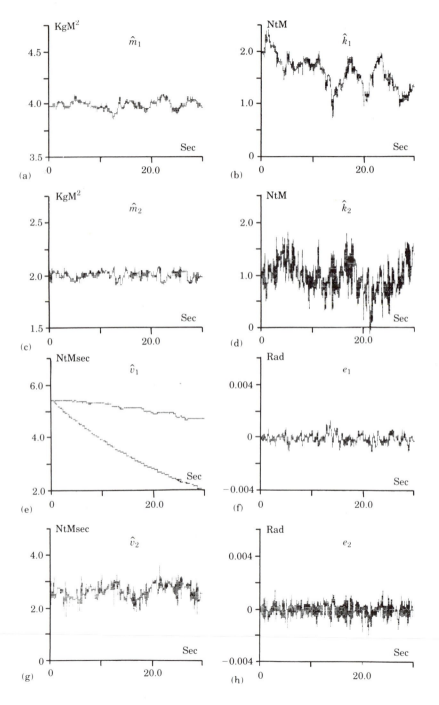

Figure 5.7 *Adaptation Too Slow to Track a Variation*

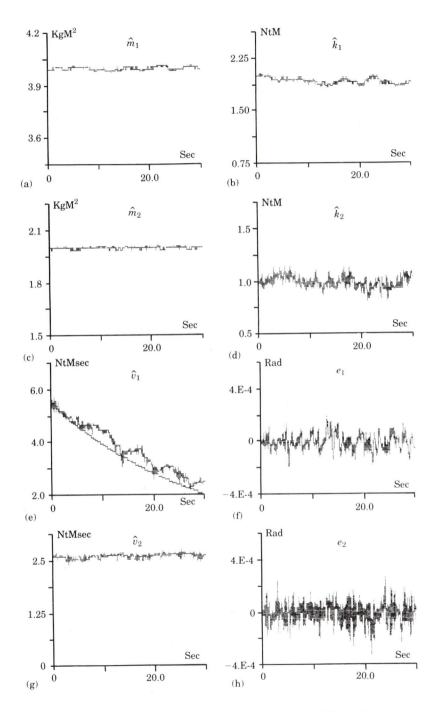

Figure 5.8 *Adaptation Tracking a Variation*

Other unmodeled dynamics tests were performed in simulation. For example, the friction coefficient of joint 1 was made to depend on the position of joint 1, although this was not known to the controller. Although these variations were too fast to be "tracked," all signals remained bounded and the system continued to operate well.

The time scale on simulations such as these may be misleading. It would have been possible to adjust the γ_i so that the adaptation was much more rapid, but an attempt was made to use numbers that were felt to be reasonable. Issues such as speed of adaptation need to be experimentally verified with an actual mechanical system that contains all the realities ignored in simulations.

5.9 Experimental Results

This section describes the results of an experimental implementation of the nonlinear model-based adaptive controller described earlier in this chapter. The implementation was done for the two major links of the Adept One, which is a four-axis "Scara"-style manipulator. The Adept One employs *direct-drive* technology in these first two links — that is, there is no gearing between the motor and the link, as in more "traditional" industrial robots. Because of this lack of gearing, and special motor design, the Adept One is perhaps the fastest industrial robot available today.

5.9.1 Dynamics of the Adept One

Figure 5.9 shows a schematic drawing of links 1 and 2 of the Adept One. Both actuators are located at the base of the robot. Joint 1's actuator applies torque τ_1 to the link 1 structure, which has rotational inertia I_1. Joint 2's actuator applies a torque τ_2 to the inner column (with inertia I_2), which drives joint 2 through a steel band. Link 2 is of mass m_2, and has its center of mass located a distance l_2 from the axis of joint 2. Link 2 has a rotational inertia about this mass center of I_3. The distance between joint axes 1 and 2 is l_1.

Figure 5.10 shows the two links as a stick figure in order to indicate the definition and positive sense of the joint angles. Due to the placement of the actuators and the transmission system, the actuator angles are not equivalent

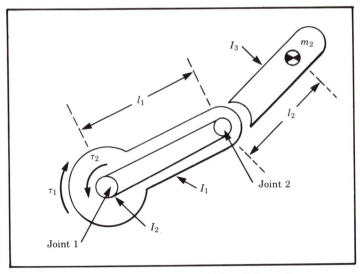

Figure 5.9 *Top View of the Adept One*

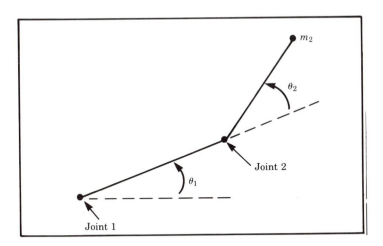

Figure 5.10 *Top View of the Adept One, Showing Angle Definitions*

to the joint angles. We will use $\Theta = [\theta_1 \; \theta_2]^T$ to indicate the joint angles, and $A = [A_1 \; A_2]^T$ to indicate the actuator angles. The transmission coupling is given by

$$\begin{bmatrix} \theta_1 \\ \theta_2 \end{bmatrix} = \begin{bmatrix} -k & 0 \\ k & k \end{bmatrix} \begin{bmatrix} A_1 \\ A_2 \end{bmatrix}, \tag{5.64}$$

where k is a constant relating encoder counts to radians.

The dynamics will be written in *actuator space*. That is, we will control (and decouple) in the two-dimensional space defined by the two actuator rotations. Hence, dynamics were developed that relate actuator accelerations and velocities to torque required at the actuators. These equations are

$$\begin{aligned} \tau_1 &= p_1 \ddot{A}_1 + p_3(-c\theta_2 \ddot{A}_2 + ks\theta_2 \dot{A}_2^2) + p_4 \dot{A}_1 + p_6 sgn(\dot{A}_1) \\ \tau_2 &= p_2 \ddot{A}_2 + p_3(-c\theta_2 \ddot{A}_1 + ks\theta_2 \dot{A}_1^2) + p_5 \dot{A}_2 + p_7 sgn(\dot{A}_2) \end{aligned}, \tag{5.65}$$

where for convenience, a dependence on θ_2 has been shown (this could obviously be written in terms of A_1 and A_2 if so desired). Terms representing viscous and Coulomb friction have been added to each joint. Actually, not all the friction exists in actuator space, but some is at the kinematic joints (i.e., in joint space). This may explain some of the perturbations we will see in the experimental results for the friction parameter estimates. The parameters p_1 through p_3 are given in terms of the physical parameters as

$$\begin{aligned} p_1 &= I_1 + m_2 l_1^2 \\ p_2 &= I_2 + I_3 + m_2 l_2^2. \\ p_3 &= m_2 l_1 l_2 \end{aligned} \tag{5.66}$$

5.9.2 Experimental Implementation

Prior to implementing the adaptive controller, rough estimates of these parameters were available as

$$\begin{aligned} p_1 &= 3.24 \\ p_2 &= 1.4 \\ p_3 &= 1.4 \end{aligned} \tag{5.67}$$

in KgM².

We need to restrict $\hat{p}_1$, $\hat{p}_2$, and $\hat{p}_3$ to lie on ranges chosen such that $\hat{M}(\Theta)$ remains invertible. However, in this case it is not sufficient to simply say $p_i > 0$. In this case it turns out that

$$Det(\hat{M}(\Theta)) = \hat{p}_1 \hat{p}_2 - \hat{p}_3^2 \cos^2 \theta_2. \tag{5.68}$$

So, rather than $\hat{p}_i > 0$, the constraint is

$$\hat{p}_1 \hat{p}_2 - \hat{p}_3^2 > 0. \tag{5.69}$$

Using the approximate knowledge of (5.67) we solve for a parameter range such that (5.69) is always true, and get

$$2.9 < \hat{p}_1 < 3.58$$
$$1.06 < \hat{p}_2 < 1.74 \tag{5.70}$$
$$1.06 < \hat{p}_3 < 1.74$$

In the case where a parameter gets stuck on its bound, the bounds can then be readjusted, but always in such a way that (5.69) is enforced.

Various servo gains were used during the experiments. Usually, gains were set with $k_p = 1000$ sec^{-2} and $k_v = 65$ sec^{-1} (for both joints), but some tests were performed with lower gains, namely, $k_p = 225$ sec^{-2} and $k_v = 30$ sec^{-1}.

Trajectories as produced by the Adept controller were used directly. These trajectories are created by interpolating smoothly a set of *via points*, which the user specifies. The nature of the interpolation used is characterized by short periods of acceleration and deceleration when path segments initiate and terminate, connected by constant velocity slews in between. Hence, for the most part, the device is experiencing accelerations only near the start or end of each individual motion, and therefore inertial parameters generally get adjusted only during these brief periods. The implementation in fixed point automatically caused a dead zone so that, usually, small amounts of noise did not cause parameters to drift.

The algorithm was written in C using 32-bit integer arithmetic running on two Motorola 68000 processors. One processor gets the desired trajectory from the usual Adept controller, and implements the computed torque servo for links 1 and 2. The second processor runs the adaptive algorithm and writes the newly updated parameter values into the first processor's memory on each servo cycle. The servo runs at a rate of 250 Hz, and each processor was utilized only about 60% of the time. Hence, it appears to be possible to control all four joints with the same hardware. If 68020 processors were available, or if the coding was carefully checked for efficiency, it would be possible to achieve a higher servo rate or use only one processor.

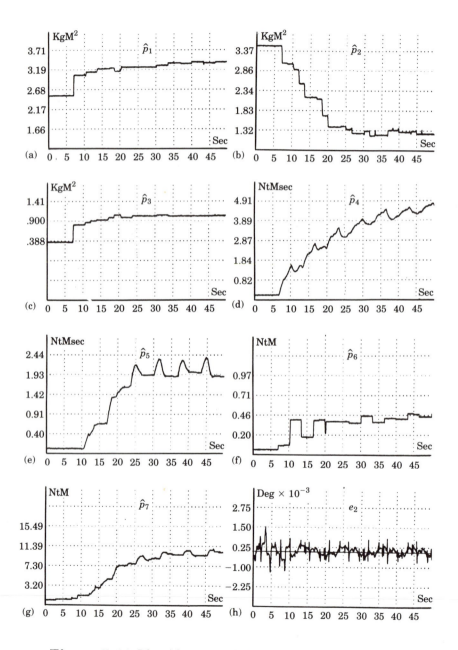

Figure 5.11 *Identification of the Adept Parameters*

	A priori	Identified
p_1	3.4 KgM^2	3.34 KgM^2
p_2	1.4 KgM^2	1.32 KgM^2
p_3	1.4 KgM^2	1.05 KgM^2

Figure 5.12 *Inertial Parameter Values Identified.*

5.9.3 Experimental Results

Figure 5.11 shows the identification of the seven parameters of the model of the two major links of the Adept One, and the angular position error for joint 2. In this experiment, the parameters were initially detuned and adaptation was enabled at $t = 7.5$ seconds. The values identified for the inertial parameters came reasonably close to those approximately known a priori by Adept (Figure 5.12).

The trajectory being followed in the identification example of Figure 5.11 is shown in actuator space in Figure 5.13. In joint space the path is given by

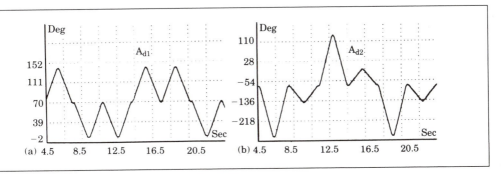

Figure 5.13 *The Desired Trajectory for Identification*

the sequence of via points:

$$[-70, \quad 0]$$
$$[0, \quad -135]$$
$$[-70, \quad 0]$$
$$[0, \quad +135]$$
$$[-70, \quad 0]$$
$$[-140, \quad 135]$$
$$[-70, \quad 0]$$
$$[-140, \quad -135]$$

Additionally, the Adept controller was set to come to rest briefly at these positions and was using half of its maximum acceleration setting and a speed setting of 125. Each cycle through this path took about 13 seconds.

It should be noted that these results can be made to appear quite different depending on how the γ_i are chosen. Namely, the adaptation can be more rapid, at the expense of more "noise" in the final value identified. In the experiment reported in Figure 5.12, since the purpose was to identify, the γ_i were set quite low, so that effectively lowpass-filtered estimates were obtained.

In Figure 5.14 partial results are shown for parameter identification using two other trajectories. The left-hand column of Figure 5.14 shows the identification of $\hat{p}_1$, $\hat{p}_2$, and $\hat{p}_3$ using a trajectory that consisted of very small motions of 15 degrees or less for each joint about the $\Theta = [0 \quad 0]$ configuration. The right-hand column of Figure 5.14 shows the identification of $\hat{p}_1$, $\hat{p}_2$, and $\hat{p}_5$ for a motion that was intended to be similar to motions that the manipulator might perform during a material transfer (or "pick and place") task. The point is that almost all trajectories are sufficient for identification, since the sufficient-excitation condition of Section 5.6 is quite easily satisfied.

Figure 5.15 shows the estimate of the inertia seen by actuator two $\hat{p}_2$ when a load is suddenly picked up by the manipulator. The mass of the load was 2.9 Kg, and was located at a distance of 0.375 m from joint 2 and so, should cause an increase in $\hat{p}_2$ of 0.409 KgM². The change in the estimate is seen to about 0.38 KgM².

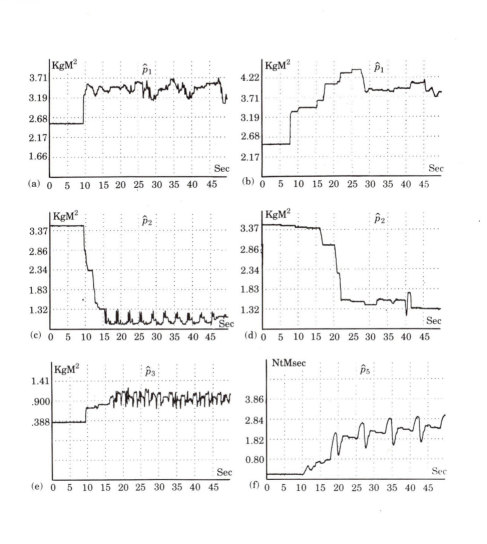

Figure 5.14 *Identification Using Other Trajectories*

Figure 5.16 shows several experiments with "fast adaptation." By fast adaptation, we mean that the γ_is were set so high that the parameter estimates changed significantly on almost every servo cycle in an apparent attempt to

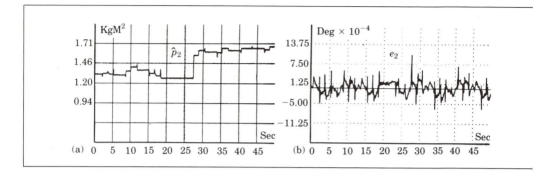

Figure 5.15 *Adaptation When Mass Is Grasped by Hand*

track unmodeled dynamics and noise. This does not lead to a useful system as far as identification is concerned. For example, see Figure 5.16g, in which $\hat{p}_5$ is seen to move quickly and often hits its a priori limits. However, with these high adaptation rates in use, performance (in terms of low-trajectory tracking errors) is maximized. Figures 5.16c and 5.16d show that the errors on the two joints greatly diminish when adaptation is enabled at $t = 23$ seconds. This performance is superior to that shown in Figure 5.12h, when the adaptation gains were set for identification. In Figures 5.16a through 5.16d the servo gains K_p and K_v were set lower to accentuate errors when the servo is mistuned. In Figure 5.16b, adaptation with high adaptation rates is enabled at $t = 12.5$ seconds, and then disabled at $t = 35$ seconds. The relative size of the servo errors in 5.16a and 5.16b indicates that no fixed setting of the controller can outperform the controller with a high adaptation rate enabled.

In Figure 5.16e the error on joint 2 is shown for the test of picking up a load at time $t = 20$ seconds using the adaptive controller with high adaptation rate and high servo gains. Figure 5.16f shows the same test (here mass is grasped at $t = 25$ seconds) with the standard Adept controller in action. These results indicate that the adaptive controller is still not outperforming Adept's fixed controller, but is quite close. However, the Adept controller is running with a servo rate of 500 Hz, as opposed to the adaptive controller's 250 Hz. We feel that with more engineering effort spent on a careful implementation, the adaptive controller will outperform the best of the fixed controllers found with present industrial controllers in certain tests. Note that tracking

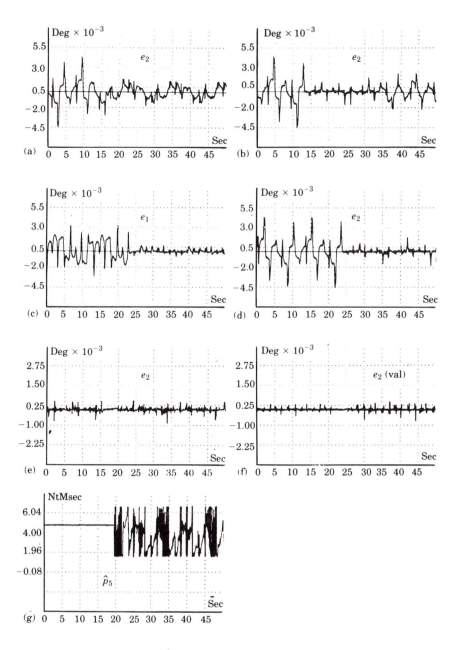

Figure 5.16 *Experiments with Fast Adaptation*

trajectories with low error is the area in which the adaptive controller should be able to improve performance relative to fixed gain controllers. However, in another important application of manipulators — fast pick and place moves (where exact trajectory shape is not important) — actuators may be saturating, and then, clearly the adaptive controller will make no improvement, and in fact, may not function as well.

Chapter 6

LEARNING CONTROL OF MANIPULATORS

6.1 Introduction

We will define *learning control* as any control scheme that improves the performance of the device being controlled as actions are repeated, and does so without the necessity of a parametric model of the system. Such schemes are advantageous when the system is difficult to model parametrically, or when the system structure is unknown. A disadvantage of such a learning controller is that generally it is applicable only in cases where a certain motion is performed over and over in cycles, so that learning can take place. The advantage of such a controller is that unstructured uncertainty can be compenstated for to the extent that it is repeatable from one trial to the next.

Our motivation in developing a learning control scheme for a mechanical manipulator came from two observations:

(1) Perhaps the major problem with parametric-model–based controllers (fixed or adaptive) is in modeling friction acting at the joints.

(2) Virtually all manipulators at work in factories today repeat their motions over and over in cycles.

Consequently, our learning controller is aimed at learning the friction effects acting at the joints and assumes that a reasonable model exists for the other dynamic effects arising from the rigid-body equations of motion.

6.2 Previous Research

Since the beginning of the research reported in this chapter several other researchers have published results in this area. It appears that Uchiyama's [138] is the only work that predates the work reported here [55, 139]. Other work on the problem of learning control applied to manipulators has been published by Arimoto et al. [140-143], Hara et al. [144], Kawamura et al. [145, 146], Mita and Kato [147], Togai and Yamano [148], Wang and Horowitz et al. [149, 150], and Atkeson and McIntyre [151].

Most researchers have adopted a similar paradigm, and the results differ mostly in analysis and approach in showing stability, convergence, and performance.

6.3 Background

As noted, the vast majority of manipulators at work in factories today repeat their motions over and over in cycles. Unfortunately, whatever errors may exist in following trajectory are also repeated from cycle to cycle. If a manipulator possesses largely repeatable dynamics, a control algorithm may be designed so that performance improves from trial to trial.

To a large extent, such trajectory-following errors are due to the simple error-driven control systems used in most present-day manipulators. The apparent solution is to increase the completeness of the dynamic model of the manipulator used in the control-law synthesis. In Chapter 2 we saw that if a perfect model of the dynamics exists, it may be used in a control law that moves the manipulator along smooth paths with zero error. These models include all the dynamic terms arising from a rigid-body model of the arm as well as friction and possibly other effects. Such models are of course never exact, with the most problematic area being that of the additional terms that attempt to model friction.

Unfortunately, the effects of friction acting at the joints of most manipulators are significant. Friction has a nonlinear dependence on joint velocity and may depend on joint position as well. Attempts at modeling the friction are further complicated by changes as the device ages, as well as variations with changes in temperature.

Since attempts to model friction will almost certainly be inaccurate, we are interested in proposing control methods in which such terms need not be modeled at all. With the lack of even a parametric model, the only plausible scheme to adapt to unmodeled effects is to repeat similar motions several times and construct feedforward torque histories that effectively cancel the unmodeled terms. Obviously, to the extent that the dynamics are not repeatable, such learning is impossible. As a practical matter, because most current robots are used in applications in which they repeat the same trajectory over and over, such schemes might find numerous practical applications.

An important goal in the design of a learning controller is to ensure the complete use of all available knowledge of the system dynamics. Intuitively, ignoring what knowledge we may have of system parameters in favor of some general learning scheme seems almost certainly to be a bad idea. We wish to model what can be modeled, and use learning for what cannot be modeled. We make direct use of what structure is known by partitioning the dynamics of the manipulator into two distinct portions: modeled and unmodeled.

The viewpoint taken in this chapter is that all terms that arise solely from the rigid-body dynamics are considered known. The effect of friction acting at the joints, which is difficult to model and may change with age and temperature, is considered unknown, lacking even a parametric model. It is possible to use the adaptive scheme of Chapter 5 to identify parameters and then use the learning scheme to refine the control further. Basically, the parametric adaptive controller identifies parameters as best it can and then these fixed parameter values are used in the modeled portion of the learning controller so that the learning controller can further correct for repeatable unmodeled dynamics over some specific task cycle.

The scheme is based on adaptively constructing feedforward torque histories for each actuator of the mechanism, which will cancel the repeatable portion of the friction effects. Since the construction of these feedforward functions is not based on a model of any kind, the learned functions may reflect almost arbitrarily complex functions originating from sources that are unknown to the designer. Compensation may even be learned for forces and torques encountered due to interaction with the environment, to the extent that it is repeatable from trial to trial.

The closer the known part of the dynamics is to the actual dynamics, the more uniform will be the stability margin (or other measure) of the distur-

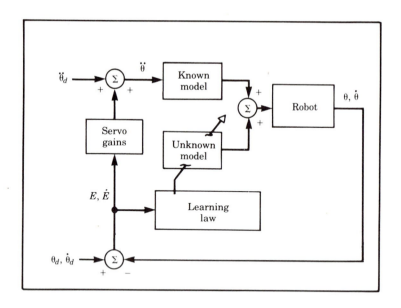

Figure 6.1 *Learning Control Scheme*

bance suppression as the configuration varies. In general, our method will cause a repeatable device to move with near-zero errors, but disturbances will be suppressed with a varying stability margin in different regions of the workspace. Servo terms in the control law can be viewed as making up a "robust" controller so that in extreme cases, performance is acceptable. This is, after all, how virtually all present-day industrial robots are controlled.

6.4 Outline of the Method

The controller consists of two parts: a control law and a learning law (see Fig. 6.1). The control law consists of the computed torque servo making use of the knowledge of the rigid-body terms and a "learned" feedforward torque function. The learning law specifies how to construct the next feedforward torque history based on the error history from the last trial. If the mechanism encounters perfectly repeatable friction as it moves, the learning law will cause a torque history to be learned that forces the device to follow the desired trajectory with zero error. Nonrepeatability, which could be viewed as a disturbance in this system, causes errors that are suppressed by the error-driven servo terms. The exact nature of disturbance suppression depends on the partitioning of the known and unknown portions of the model in the control law.

The form of learning law proposed here is that of a linear filter. That is, the entire function of servo errors observed in the previous trial is filtered. The output of this filtering operation is a function that is added to the previous torque history to generate the next torque history. Although the filter is linear, through the interactions with the nonlinear equations of motion of the mechanism, complex nonlinear torque functions are learned after several trials.

Analysis of the entire closed-loop system yields a criterion to be met for convergence. This criterion may be met by proper choice of the filter coefficients in the learning law. If the criterion is met, the torque history will approach the particular history that causes the device to move along the trajectory with zero error. In the presence of nonrepeatability or of constant disturbances the system lowers the error to some nonzero limit imposed by the nonrepeatability.

6.5 The Control Law and Error Equation

The control law is proposed as

$$ T = \hat{M}(\Theta)\left[\ddot{\Theta}_d + K_v\dot{E} + K_p E\right] + \hat{V}(\Theta, \dot{\Theta}) + \hat{G}(\Theta) + \hat{F}_k, \qquad (6.1) $$

where $\hat{F}_k$ is the feedforward torque function after trial k.

On the first cycle, the feedforward torque function is zero. After each trial, it is adjusted according to a learning law (see following section).

By combining (2.1) and (6.1) we write the system dynamics in error space as

$$ \ddot{E} + K_v\dot{E} + K_p E = \hat{M}^{-1}(\Theta)\left[F(\dot{\Theta}) + T_d - \hat{F}_k\right], \qquad (6.2) $$

or more compactly as

$$ \ddot{E} + K_v\dot{E} + K_p E = D(\Theta, \dot{\Theta}) - \hat{D}_k. \qquad (6.3) $$

Our analysis of the learning controller is based on the assumption that since the modeled portion of the dynamics results in fairly good control, the errors are small. Also, we assume that $D(\Theta, \dot{\Theta})$ is largely repeatable from trial to trial. Hence, we see that (6.3) may be written as n equations (one per joint) in the form

$$\ddot{e} + k_v \dot{e} + k_p e = d(t) - \hat{d}_k, \tag{6.4}$$

where $d(t)$ is the ith element of $\hat{M}^{-1}(\Theta)\left[F(\dot{\Theta}) + T_d\right]$ and is considered a fixed function of time. Likewise, $\hat{d}_k$ is the ith element of $\hat{M}^{-1}(\Theta)\hat{F}_k$. Note that to avoid multiple subscripts we have dropped the usual i subscript indicating the ith joint in (6.4).

6.6 Convergence

We wish to design a learning scheme in which the feedforward torque function $\hat{d}_k$ converges to a function that causes the trajectory to be followed with small error, ideally $d(t)$.

The form of the learning law for each joint is proposed as

$$\hat{d}_{k+1} = \hat{d}_k + P \star e_k, \tag{6.5}$$

where P is the impulse response of a linear filter, $\star$ is the convolution operator, and subscript k refers to the trial number. Laplace transforming (6.5) yields

$$\hat{D}_{k+1}(s) = D_k(s) + P(s)e_k(s). \tag{6.6}$$

Using the Laplace transform of (6.4) and substituting into (6.6) yields

$$\hat{D}_{k+1}(s) = \hat{D}_k(s) + P(s)H(s)\left[D(s) - \hat{D}_k(s)\right], \tag{6.7}$$

where

$$H(s) = \frac{1}{s^2 + k_v s + k_p}. \tag{6.8}$$

Collecting terms,

$$\hat{D}_{k+1}(s) = [1 - P(s)H(s)]\,\hat{D}_k(s) + P(s)H(s)D(s), \tag{6.9}$$

which is a recursion in the transform of the feedforward torque function. The solution to this recursion is seen to be

$$\hat{D}_k(s) = D(s) + CG^k(s), \tag{6.10}$$

where $G(s) = 1 - P(s)H(s)$ and C is a constant.

The two questions we ask are Under what conditions does (6.10) converge? and, If it converges, what does it converge to?

It is clear that if the second term in (6.10) goes to zero with increasing trial number, then the recursion converges, and indeed converges to the transform of the unknown dynamic function along the desired trajectory.

The recursive nature of (6.9) might be thought of instead as an infinite sequence of filtering operations. In an infinite sequence of identical filters, the frequency response of each filter must be less than one for all frequencies. If it is greater than one for any particular frequency, that frequency component, if excited, will grow without bound. Thus $G(s)$ must have a frequency response less than one for all frequencies. A somewhat more formal proof follows.

Let

$$g_k(t) = \mathcal{L}^{-1}\left[G^k(s)\right], \tag{6.11}$$

where $\mathcal{L}^{-1}$ is the inverse Laplace operator. Then, by Parseval's relation:

$$\int_{-\infty}^{+\infty} g_k^2(t)dt = \frac{1}{2\pi} \int_{-\infty}^{+\infty} \left|G^k(j\omega)\right|^2 d\omega \tag{6.12}$$

and

$$\left|G^k(j\omega)\right| = \left|G(j\omega)\right|^k. \tag{6.13}$$

Then

$$\int_{-\infty}^{+\infty} g_k^2(t)dt = \frac{1}{2\pi} \int_{-\infty}^{+\infty} \left|G(j\omega)\right|^{2k} d\omega. \tag{6.14}$$

If

$$\left|G(j\omega)\right| < 1 \quad \forall \omega \quad \text{and} \quad \lim_{\omega \to \infty} \left|G(j\omega)\right| = 0 \tag{6.15}$$

then

$$\lim_{k \to \infty} \left|g_k(t)\right| = 0. \tag{6.16}$$

Given (6.15), the right side of (6.14) equals zero due to the monotone convergence theorem [54]. The second condition in (6.15) ensures that the convergence theorem can be extended to the infinite interval case. Therefore given the condition stated in (6.15) we must have

$$\lim_{k \to \infty} G^k(s) = 0. \tag{6.17}$$

Note that the filter used in the learning law appears as part of this critical transfer function:

$$G(s) = 1 - P(s)H(s) = \frac{s^2 + k_v s + k_p - P(s)}{s^2 + k_v s + k_p}. \tag{6.18}$$

We choose the form of the learning filter to be

$$P(s) = s^2 + (k_v - \mu)s + (k_p - \mu), \tag{6.19}$$

where μ is a constant.

Any reasonable implementation of this filter will be noncausal. However, because we have an entire history of errors to filter from the previous trial, a noncausal filter may be implemented easily.

From (6.18) and (6.19):

$$G(s) = \frac{\mu(s+1)}{s^2 + k_v s + k_p}. \tag{6.20}$$

Assuming gains were chosen such that k_v and k_p form an approximately critically damped system, the magnitude of (6.20) along the imaginary axis is

$$|G(j\omega)| = \mu \frac{\sqrt{\omega^2 + 1}}{\omega^2 + \frac{k_v^2}{4}}. \tag{6.21}$$

The maximum of (6.21) occurs at

$$\omega_{max} = \sqrt{\frac{k_v^2}{4} - 2}, \tag{6.22}$$

where the magnitude is

$$|G(j\omega_{max})| = \mu \frac{\sqrt{\frac{k_v^2}{4} - 1}}{\frac{k_v^2}{2} - 2}. \tag{6.23}$$

In order to keep (6.21) less than one over all frequencies, we see we must choose

$$0 < \mu < \frac{\frac{k_v^2}{2} - 2}{\sqrt{\frac{k_v^2}{4} - 1}}. \tag{6.24}$$

In choosing the filter, we must be sure that the frequency response of (6.21) is less than one at all frequencies. We might also choose a more complicated filter than (6.19), which would allow us to meet the frequency-response criteria while also allowing more design freedom. This freedom might be used to ensure that a filter with a sharp cutoff frequency is implemented. Such a lowpass characteristic may in fact be necessary due to unmodeled vibrations in the mechanism.

6.7 Simulation Results

The implementation of the system is in discrete time. Although a discrete analysis has been done for the system, it lends no particular insight to the problem, and so the implementation may be considered simply as discrete equivalents of the continuous design.

The discrete version of the control law is

$$
\begin{aligned}
T(n) = & \hat{M}(\Theta(n)) \left[\ddot{\Theta}_d(n) + \alpha e(n) + \beta e(n-1) \right] \\
& + \hat{V}(\Theta(n), \dot{\Theta}(n)) + \hat{G}(\Theta(n)) + \hat{F}_k(n).
\end{aligned} \tag{6.25}
$$

The form of the learning filter was found by simply forming a discrete equivalent of the continuous filter of (6.19). In discretizing we used the relations

$$s = \frac{1}{2\Delta T}(z - z^{-1}) \tag{6.26}$$

and

$$s^2 = \frac{1}{\Delta T^2}(z - 2 - z^{-1}), \tag{6.27}$$

where ΔT is the sample period.

The torque histories are stored in arrays in memory. Memory requirements are not large by today's standards. For example, a six-jointed arm being servoed at 100 Hz over a cyclic path of 5 seconds' duration would require 3000 stored values.

A two-degree-of-freedom arm is sufficient to observe any coupling effects that might affect the algorithm. As in other simulations in this thesis, a two-jointed planar arm was simulated. The manipulator is as pictured in Figure 3.3 and the dynamics are given by (3.42).

The control law (6.25) was implemented with a sample period of 0.01 second and gains computed to yield approximate critical damping with a stiffness of 200 NtM/Rad:

$$\begin{aligned}
\alpha &= \begin{bmatrix} 3028 & 0 \\ 0 & 3028 \end{bmatrix} \\
\beta &= \begin{bmatrix} -2828 & 0 \\ 0 & -2828 \end{bmatrix}.
\end{aligned} \tag{6.28}$$

The desired trajectory can be arbitrarily complex as long as the limits on the actuators are not exceeded. For the experiments reported here, a simple cyclic path with continuous position, velocity, and acceleration was used. This path is defined by specifying three points in space, and connecting them with three cubic splines. A quintic spline segment is used to move from rest onto the cyclic path, and another is used to decelerate to rest after any number of cycles are performed. The entire path has a parameter of time that can be scaled to change the speed of the path. In these experiments a cycle lasted 2.4 seconds and the cycle was defined by the three path points specified in Figure 6.2.

Figure 6.3 shows one cycle of the desired trajectory of the end-effector in the X–Y plane.

Although the analysis was performed assuming that the model-based portion of the controller was accurate enough that the arm was decoupled, in

Point	$X(M)$	$Y(M)$	Θ_1	Θ_2
1	0.258	−0.023	70°	−150°
2	0.633	0.112	60°	−100°
3	0.953	0.211	25°	−25°

Figure 6.2 *Trajectory via Points*

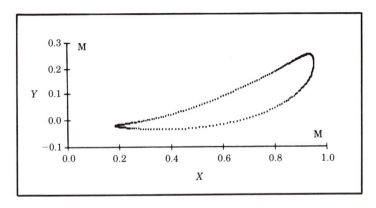

Figure 6.3 *Desired Spatial Trajectory*

the simulations very large errors were introduced in the supposedly known parameters. All mass parameters were assumed to be known only within 25%, and so values that had a 25% error were chosen for m_1 and m_2. Furthermore, the velocity terms and the gravity terms were not computed at all in the controller, i.e., $\hat{V}(\Theta, \dot{\Theta}) = 0$ and $\hat{G}(\Theta) = 0$. Thus the torque function learned represents the effects of velocity terms, gravity terms, friction terms, plus approximately 25% of the inertial torques as well. When knowledge of the mass constants is better, and/or when velocity and gravity terms are computed, performance is even better than the results shown below.

Figures 6.4 and 6.5 show how the error is reduced with the number of trials. Here, the error value is calculated as the average error magnitude over an entire cycle. Note the use of a log scale and units of degrees for the average error values.

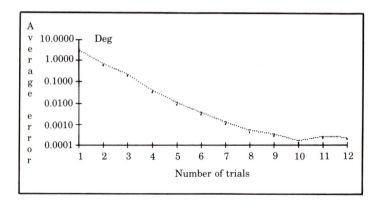

Figure 6.4 *Learning Curve for Joint 1*

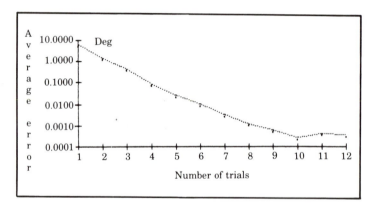

Figure 6.5 *Learning Curve for Joint 2*

Figures 6.6 and 6.7 show the feedforward torque functions that were learned after 10 trials. Because the simulation is perfectly repeatable, the results are extremely good.

Figures 6.8 and 6.9 show the learning curves for the case when random torque noise at the joints is added to simulate some external disturbance or nonrepeatability. The errors are on a log scale in units of degrees. The noise was evenly distributed between −2.0 and 2.0 NtM at each joint.

Figures 6.10 and 6.11 show the feedforward functions that were learned after 10 trials. Actually, feedforward functions learned after only one or two trials are quite good.

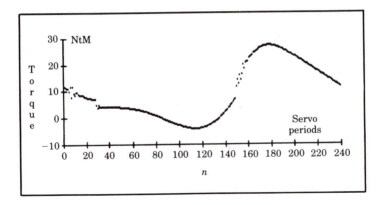

Figure 6.6 *Learned Torque Function for Joint 1*

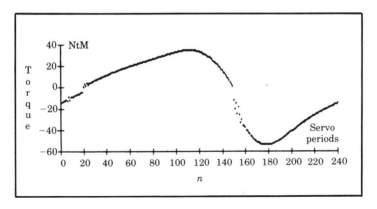

Figure 6.7 *Learned Torque Function for Joint 2*

Future work will include implementation on an actual manipulator for testing. The scheme has been run on a single joint of a PUMA 560 with good results [55].

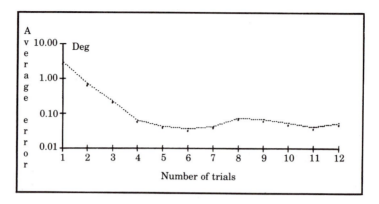

Figure 6.8 *Learning Curve for Joint 1 (with Noise)*

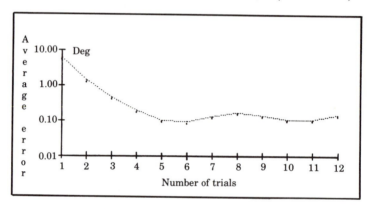

Figure 6.9 *Learning Curve for Joint 2 (with Noise)*

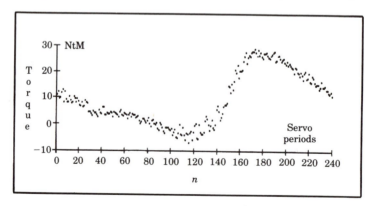

Figure 6.10 *Learned Torque Function for Joint 1 (with Noise)*

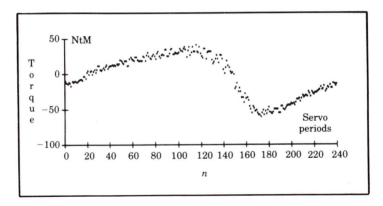

Figure 6.11 *Learned Torque Function for Joint 2 (with Noise)*

Chapter 7

CONCLUSIONS

7.1 Robustness

A robustness result for the computed torque servo has been developed. It gives conditions, which are a function of the uncertainty in parameter estimates, such that the nonlinear model-based controller will be stable. The result is a sufficient-only theorem, and is felt to be quite conservative, hence, a much stronger conjecture has been stated. The proof of this conjecture remains an open problem.

7.2 Adaptive Control

A globally stable adaptive-control scheme for a complex nonlinear system has been designed. The adaptation process can be added to the model-based control formulation for robot manipulators (sometimes called the computed torque method) without otherwise altering the controller structure.

Using the results in Section 5.6, trajectories especially well suited to identification of parameters could be preplanned. However, most real-world trajectories carried out by industrial robots are sufficiently exciting, so the scheme could be used as an on-line controller.

This method could be directly applied to a Cartesian (or *operational space*) based control scheme such as the one reported in Khatib [41]. This would be essentially straightforward with minor additions to ensure that the product of W and the Cartesian mass matrix remain bounded at all times. An interesting application would be to redundant manipulators controlled in

Cartesian space. For such a manipulator, internal motions of the manipulator that do not affect the end point might be designed to be well suited to identifying dynamic properties of the manipulator. The method might also be extended to include an active force control servo, where some identified parameters are associated with properties of the task surfaces rather than strictly with the manipulator itself. On the theoretical side, further research and analysis of coping with a more general class of unmodeled dynamics is required, as is presently the case in all adaptive control work.

It is important to realize that although a significant amount of computation is required to implement the controller, it is not so much as to make the method impractical. This has been shown (albeit in a simple case) in Section 5.9 for a two-link industrial robot. There have been implementations of the computed torque servo running in several laboratories [24, 41] for some time, and the adaptive controller does not require that much more computation. To make a rough estimate, we feel that N (the number of joints) processors of approximately the power of the Motorola 68000 would be more than sufficient to implement the algorithm on a general manipulator. Given an industrial robot controller that already must provide power supplies, a backplane, etc., the incremental cost of adding such a processor is probably presently about $500, and still dropping. On the other hand, the entire system of manipulator and controller may cost $50,000 or more. It is time to drop some of the engineering attitudes of two decades ago, and start using this computational power.

7.3 Learning Control

A learning scheme has been developed for the case of repeated trials of a path. The primary benefit of the approach is that it allows compensation for effects such as friction, which are otherwise difficult to model. A secondary advantage of the scheme is that it can also adapt to torques due to velocity and gravity effects. After sufficient learning, arbitrary paths that are within the torque and velocity limits of the arm can be executed with errors lower bounded only by the nonrepeatability of the manipulator.

Another interesting application of our scheme would be as a method of verification of dynamic models. If a "complete" dynamic model of the manipulator is used in the modeled portion of the control, the feedforward torque learned should be zero. Thus presence of a nonzero learned torque function indicates that the model is not exact. If the torque function is small or zero over most of the trajectory, the proposed model is proven accurate.

Due to the inevitable presence of unmodeled dynamics, this learning scheme might find use in compensating for the residual unmodeled effects after a parameter-identification scheme is used. First the adaptive controller of Chapter 5 would be used to identify unknown parameters in a proposed model. Once so identified, the adaptation portion of the controller could be removed, and the fixed model-based controller run using the learning algorithm of Chapter 6 to further reduce trajectory-following errors.

Appendix A

NORMS AND NORMED SPACES

A.1 Introduction

A **norm** is a function whose range is $\mathcal{R}^A+$ (the positive reals), which, roughly speaking, indicates the "size" of some quantity. Among the quantities for which a norm might be defined are vectors, matrices, functions of time, and operator gains. Norms can play an important role in stability theory by providing a measure of the size of various quantities. Here we give a brief introduction to norms. For more details see Desoer and Vidyasagar [48], and Vidyasagar [51].

A.2 Vector Norms

We will consider norms defined on $\mathcal{R}^n$. Let the zero vector in this space be denoted by 0 (same symbol as scalar zero [let context differentiate them]). We say that the function $\|\cdot\| : \mathcal{R}^n \to \mathcal{R}^A+$ is a **norm** on $\mathcal{R}^n$ if and only if

(1) $x \in \mathcal{R}^n$ and $x \neq 0 \Rightarrow \|x\| > 0$.

(2) $\|\alpha x\| = |\alpha| \|x\|, \quad \forall \alpha \in \mathcal{R}, \quad \forall x \in \mathcal{R}^n$.

(3) $\|x + y\| \leq \|x\| + \|y\|, \quad \forall x, y \in \mathcal{R}^n$.

Many functions $\|\cdot\|$ can be designed with these properties. Given a particular norm function $\|\cdot\|_p$ the pair $(\mathcal{R}^n, \|\cdot\|_p)$ is called a **normed space**.

104

The following functions can be shown to satisfy the above conditions (see Atkinson [52]) and hence may serve as norms:

$$\|x\|_1 \triangleq \sum_{i=1}^{n} |x_i|;$$

$$\|x\|_p \triangleq \left(\sum_{i=1}^{n} |x_i|^p \right)^{\frac{1}{p}}, \quad p \in [1, \infty); \qquad (A.1)$$

$$\|x\|_\infty \triangleq \max_i |x_i|.$$

Note that $\|x\|_2$ is the usual Euclidean norm of x:

$$\|x\|_2 = \sqrt{x_1^2 + x_2^2 + \ldots + x_n^2}. \qquad (A.2)$$

An important result (see Desoer and Vidyasagar [48]) is that all norms are equivalent in the sense that there exist positive constants α_1 and α_2 such that

$$\alpha_1 \|x\|_a \le \|x\|_b \le \alpha_2 \|x\|_a, \quad \forall x \in \mathcal{R}^n. \qquad (A.3)$$

This means that any norm can be used to show boundedness or convergence since all norms define the same topology on $\mathcal{R}^n$. Hence, often the symbol $\| \cdot \|$ will be used without specifying exactly which norm is meant.

A.3 Induced Matrix Norms

In the same way in which a norm can be used to measure the "size" of a vector, norms can be defined for matrices as well. For a matrix $A \in \mathcal{R}^{m \times n}$ we define

$$\|A\|_p \triangleq \sup_{x \neq 0} \frac{\|Ax\|_p}{\|x\|_p} \quad \forall x \in \mathcal{R}^n. \qquad (A.4)$$

Because the definition of the matrix norm depends on the choice of vector norm, $\|A\|_p$ is called the **induced matrix norm** of the vector norm $\|x\|_p$.

It can be shown that for such an induced norm

$$\|AB\|_p \le \|A\|_p \|B\|_p, \quad \forall A, B \in \mathcal{R}^{n \times n}$$
$$\|Ax\|_p \le \|A\|_p \|x\|_p, \quad \forall A \in \mathcal{R}^{n \times n}, x \in \mathcal{R}^n. \qquad (A.5)$$

A useful result (see Atkinson [52]) for $p = 2$ is

$$\|A\|_2 = \sqrt{\max_i \lambda_i(A^T A)}, \tag{A.6}$$

where $\lambda_i(A^T A)$ is the ith eigenvalue of $A^T A$. If A is symmetric then

$$\|A\|_2 = \max_i \lambda_i(A). \tag{A.7}$$

A.4 Function Norms

Just as a "size" can be associated with a vector or a matrix, we can also define the norm of a function. Consider a function $f : R^A+ \to R$ and associate with it the norm

$$\|f\|_p \triangleq \left(\int_0^\infty |f(t)|^p dt \right)^{\frac{1}{p}}. \tag{A.8}$$

If the above integral exists (i.e., is finite) then we say that the function $f : R^A+ \to R$ belongs to the subspace L_p of the space of all possible functions.

Likewise, we can define a norm for vector-valued functions. For example, considering a function $g : R^A+ \to R^n$, a family of norms is given by

$$\|g\|_p \triangleq \left(\int_0^\infty \|g(t)\|_p^p dt \right)^{\frac{1}{p}}. \tag{A.9}$$

If the above integral exists (i.e., is finite) then we say that the function $g : R^A+ \to R^n$ belongs to the subspace L_p^n of the space of all possible vector-valued functions.

It is well known that the linear space L_p together with the norm $\|\cdot\|_p$ forms a normed space that is **complete** (i.e., every Cauchy sequence in L_p converges to a unique element in L_p). Such a complete normed space is called a **Banach space**.

Associated with the normed space L_p is an *extended space* L_{pe} with $L_p \subset L_{pe}$. In addition to the class of functions in L_p this extended space also includes truncated functions of the form:

$$f_T(t) = f(t) \quad t \leq T$$
$$= 0 \quad t > T, \tag{A.10}$$

such that

$$\|f_T\|_p < \infty \quad \forall T. \tag{A.11}$$

The idea is that there are some functions that if not truncated do not belong to L_p, but the associated truncated function will belong to L_p for any finite T. This definition of L_{pe} facilitates consideraton of systems in which subsystems may be unstable but the entire system is stable. For convenience we will sometimes use the notation $\|f\|_{Tp}$ to mean $\|f_T\|_p$.

A.5 Operator Gains

Let $\mathcal{H}$ be a causal operator $\mathcal{H} : f \mapsto g$. The L_p gain of $\mathcal{H}$ is denoted $\|\mathcal{H}\|_p$ and is defined as the smallest value of γ such that

$$\|g\|_p = \|\mathcal{H}f\|_p \le \gamma\|f\|_p + \beta \quad \forall f \in L_p. \tag{A.12}$$

Loosely speaking, the L_p gain measures the amplification of the operator $\mathcal{H}$.

A special result for $p = 2$ is (see Desoer and Vidyasagar [48]) that for a convolution-type operator $\mathcal{H} : f \mapsto g$ given by

$$g(t) = \int_{-\infty}^{\infty} h(t - \tau)f(\tau)d\tau, \quad \forall t \in \mathcal{R}, \tag{A.13}$$

the L_2 gain is given by

$$\|\mathcal{H}\|_2 = \max_{\omega \in \mathcal{R}} |\hat{h}(j\omega)|, \tag{A.14}$$

where $\hat{h}(s)$ is the Laplace transform of the impulse response of $\mathcal{H}$.

A.6 L_∞ Function Norms and Operator Gains

In Chapter 3 we made use of a normed space usually denoted L_∞. In this section we give the defintions of the L_∞ function norms and gains used throughout Chapter 3.

The L_∞ norm associated with a scalar function of time $a : \mathcal{R}^A+ \to \mathcal{R}$ is

$$\|a\|_\infty \triangleq \sup_t |a(t)|, \tag{A.15}$$

and so, if $\|a\|_\infty < \infty$, we say that $a \in L_\infty$.

The norm associated with a vector function of time $x : \mathcal{R}^A+ \to \mathcal{R}^n$ is

$$\|x\|_\infty \overset{\triangle}{=} \max_i \ \sup_t |x_i(t)|, \tag{A.16}$$

and so, if $\|x\|_\infty < \infty$, we say that $x \in L_\infty^n$. For vector-valued functions that may eventually grow to infinity in magnitude, we will define

$$\|x\|_{T\infty} \overset{\triangle}{=} \max_i \ \sup_{t<T} |x_i(t)|, \tag{A.17}$$

and such a function is said to belong to $L_{\infty e}^n$.

The induced norm associated with a matrix function of time $A : \mathcal{R}^A+ \to \mathcal{R}^{n \times n}$ is

$$\|A\|_{i\infty} \overset{\triangle}{=} \max_i \ \sup_t \sum_j |a_{ij}(t)|. \tag{A.18}$$

Consider the convolution operator $\mathcal{H} : a \mapsto b$, such that

$$b(t) = \int_{-\infty}^{\infty} h(t - \tau)a(\tau)d\tau, \quad \forall t \in \mathcal{R}, \tag{A.19}$$

The L_∞ gain of this operator (see Desoer and Vidyasagar [48]) is given by

$$\|\mathcal{H}\|_\infty = \int_0^{\infty} |h(t)|dt. \tag{A.20}$$

Appendix B

LYAPUNOV
STABILITY THEORY

B.1 Introduction

In this appendix we quickly review Lyapunov stability theory [49, 50]. In particular, we review what is called Lyapunov's *second* or *direct* method. It is one of the few techniques that can be applied directly to nonlinear systems to investigate their stability. Other related techniques with application to nonlinear systems are the small-gain theorem and passivity [48]. As a means of quickly introducing Lyapunov's method (in sufficient detail for our needs) we will give an extremely brief introduction to the theory, and then proceed directly to several examples.

B.2 Lyapunov's Direct Method

Lyapunov's method is concerned with determining the stability of a differential equation

$$\dot{X} = f(X), \qquad (B.1)$$

where X is $m \times 1$ and $f(\cdot)$ may be nonlinear. To prove a system stable by Lyapunov's second method, you must propose a generalized energy function $v(X)$ that has the following properties.

(1) $v(X)$ has continuous first partial derivatives and $v(X) > 0$ for all X except $v(0) = 0$.

(2) $\dot{v}(X) \leq 0$. Here, $\dot{v}(X)$ means the change in $v(X)$ along all system trajectories.

These properties may hold only in a certain region or may be global, with correspondingly weaker or stronger stability results. The intuitive idea is that a positive definite "energy-like" function of state is shown to always decrease or remain constant — hence, the system is stable in the sense that the size of the state vector is bounded.

B.3 LaSalle's Extension: Invariant Sets

When $\dot{v}(X)$ is strictly less than zero, asymptotic convergence of the state to the zero vector can be concluded. Lyapunov's original work was extended in an important way by LaSalle and Lefschetz [50], who showed that in certain situations, even when $\dot{v}(X) \leq 0$ (note equality included), asymptotic stability may be shown. For this result we introduce the notion of an *invariant set*.

An invariant set G is characterized by the property that if a point x_0 is in G, then the entire system trajectory passing through x_0 lies in G [50]. Thus, for example, any trajectory that forms a closed path is an invariant set.

Let $v(X)$ be a scalar function with continuous partial derivatives. Let Ω_1 designate the region where $v(X) < v^*$. Assume that Ω_1 is bounded and that within Ω_1

$v(X) > 0$ for $x \neq 0$,

$\dot{v}(X) \leq 0$.

Let R be the set of all points within Ω_1, where $\dot{v}(X) = 0$, and let M be the largest invariant set in R. Then every solution $x(t)$ in Ω_1 tends to M as $t \to \infty$.

B.4 Illustrative Examples

These examples are intended to introduce the method of Lyapunov stability analysis.

Example B.1

Consider the linear system

$$\dot{X} = -AX, \qquad (B.2)$$

where A is $m \times m$ and positive definite. Propose the *Lyapunov candidate function*

$$v(X) = \frac{1}{2}X^T X, \qquad (B.3)$$

which is continuous and everywhere nonnegative. Differentiating yields

$$\begin{aligned} \dot{v}(X) &= X^T \dot{X} \\ &= X^T(-AX) \qquad (B.4) \\ &= -X^T AX, \end{aligned}$$

which is everywhere nonpositive. Hence, (B.3) is indeed a Lyapunov function for the system of (B.2). The system is asymptotically stable because $\dot{v}(X)$ can be zero only at $X = 0$; everywhere else X must decrease.

Example B.2

Consider the linear system driven by a bounded disturbance

$$\dot{X} = -AX + \nu, \qquad (B.5)$$

where A is $m \times m$ and positive definite and ν is $m \times 1$ and bounded as

$$\|\nu\| < \nu_{max}. \qquad (B.6)$$

Propose the Lyapunov candidate function

$$v(X) = \frac{1}{2}X^T X, \qquad (B.7)$$

which is continuous and everywhere nonnegative. Differentiating yields

$$\begin{aligned} \dot{v}(X) &= X^T \dot{X} \\ &= X^T(-AX + \nu) \qquad (B.8) \\ &= -X^T AX + X^T \nu. \end{aligned}$$

Note that the second term in (B.8) is of indeterminant sign, but for sufficiently large X, the first term will dominate and ensure a negative $\dot{v}(X)$. To be specific, we see that in the region given by

$$X > A^{-1}\nu, \tag{B.9}$$

we have $\dot{v}(X) < 0$. The proper interpretation of this (also including nonzero initial state) is that X is bounded and will *eventually* be confined to the region

$$X < A^{-1}\nu. \tag{B.10}$$

This is consistent with the intuition of the situation of a bounded signal driving an asymptotically stable system.

Example B.3

Consider a mechanical spring-damper system in which both the spring and damper are nonlinear:

$$\ddot{x} + b(\dot{x}) + k(x) = 0. \tag{B.11}$$

The functions $b(\cdot)$ and $k(\cdot)$ are first- and third-quadrant continuous functions such that

$$\begin{aligned} \dot{x}b(\dot{x}) &> 0 \quad for \quad x \neq 0, \\ xk(x) &> 0 \quad for \quad x \neq 0. \end{aligned} \tag{B.12}$$

Proposing the Lyapunov candidate function

$$v(x, \dot{x}) = \frac{1}{2}\dot{x}^2 + \int_0^x k(\lambda)d\lambda, \tag{B.13}$$

we are led to

$$\begin{aligned} \dot{v}(x, \dot{x}) &= \dot{x}\ddot{x} + k(x)\dot{x} \\ &= -\dot{x}b(\dot{x}) - k(x)\dot{x} + k(x)\dot{x} \\ &= -\dot{x}b(\dot{x}). \end{aligned} \tag{B.14}$$

Hence, $\dot{v}(\cdot)$ is nonpositive, but is only semidefinite, since it is not a function of x but only of $\dot{x}$. In order to conclude asymptotic stability we have to ensure that it is not possible for the system to "get stuck" with nonzero x. To study all trajectories for which $\dot{x} = 0$, we must consider

$$\ddot{x} = -k(x) \tag{B.15}$$

for which $x = 0$ is the only (and hence largest) invariant set. Hence the system will only come to rest if $x = \dot{x} = \ddot{x} = 0$.

Appendix C

STRICTLY POSITIVE
REAL SYSTEMS

C.1 Introduction

Here we briefly review positive real transfer functions. For more information, see references [48] and [64].

C.2 Frequency-Domain Definition

A rational transfer function $H(s)$ is called strictly positive real (SPR) if

(1) It has no zeros in the right half-plane, and

(2) Evaluated along the imaginary axis, the real part of $H(s)$ is always positive.

Furthermore, from condition (2) above, we can conclude that the phase shift of $H(s)$ must always have a magnitude of 90 degrees or less, and so the number of poles can exceed the number of finite zeros at most by one.

C.3 Engineering Definition

An SPR transfer function corresponds one to one with a passive network, for example, a connection strictly composed of inductors, capacitors, and resistors [64]. Imagine that $H(s)$ corresponds to the impedance of a passive one-port, that is,

$$\frac{v(s)}{i(s)} = H(s). \tag{C.1}$$

Then clearly, if the one-port is passive, the real part of the impedance must be positive at all frequencies. If it were negative at some frequency, the one-port would be supplying power at that frequency, and therefore could not be composed soley of passive elements.

C.4 Time-Domain Definition

Considering the example of a one-port passive network, with current viewed as the input and voltage as the output, the following time-domain definition of a passive system is clear.

A system with input $i(t)$ and output $v(t)$ is passive if (assuming it contained no energy when assembled at $t = -\infty$) [48]:

$$\int_{-\infty}^{t} v^T(t)i(t)dt > 0 \quad \forall t. \tag{C.2}$$

That is, the power consumed at any time must be positive. Note that in (C.2), the input and output may both be $m \times 1$ vectors.

C.5 State–Space Definition

This property of SPR systems is sometimes called *the SPR lemma* [64]: If $H(s)$ is SPR, then the state space realization with state x,

$$\begin{aligned} \dot{x} &= Ax + Bu, \\ y &= Cx, \end{aligned} \tag{C.3}$$

has the following property. There exists two positive definite matrices $P > 0$ and $Q > 0$ such that

$$\begin{aligned} A^T P + PA &= -Q \\ PB &= C^T. \end{aligned} \tag{C.4}$$

REFERENCES

References for History, Kinematics, and Path Planning

[1] R.C. Goertz and F. Bevilacqua. "A Force-Reflecting Positional Servomechanism." *Nucleonics* 1952; 14:43–55.

[2] J.F. Engelberger. *Robotics in Practice.* New York: Amacom, 1980.

[3] H.A. Ernst. "MH-1, A Computer-Oriented Mechanical Hand." Proceedings of the 1962 Spring Joint Computer Conference, San Francisco, 1962.

[4] J. McCarthy et al. "A Computer with Hands, Eyes, and Ears." 1968 Fall Joint Computer Conference, AFIPS Proceedings, 1968:329–338.

[5] V.D. Scheinman. "Design of a Computer Manipulator." Stanford, Calif.: Stanford Artificial Intelligence Laboratory Memo AIM-92, 1969.

[6] L.G. Roberts. "Homogeneous Matrix Representation and Manipulation of N-Dimensional Constructs." MIT Lincoln Labs, Document MS-1045, May 1965.

[7] J. Denavit and R.S. Hartenberg. "A Kinematic Notation for Lower-Pair Mechanisms Based on Matrices." *J. Appl. Mechanics* 1955.

[8] D. L. Peiper. "The Kinematics of Manipulators under Computer Control." Stanford, Calif.: Stanford Artificial Intelligence Laboratory Memo AIM-72, 1968.

[9] R. P. Paul. "Modeling, Trajectory Calculation, and Servoing of a Computer Controlled Arm." Stanford, Calif.: Stanford Artificial Intelligence Laboratory Memo AIM-177, 1972.

[10] R. Paul. "Cartesian Coordinate Control of Robots in Joint Coordinates." Proceedings of the 3rd IFTOMM International Symposium on Theory and Practice of Robots and Manipulators, Udine, 1978.

[11] D. E. Whitney. "Resolved Motion Rate Control of Manipulators and Human Prothesis." *IEEE Trans. on M.M.S.* 1969; 10(2).

[12] M. Kahn and B. Roth. "The Near Minimum Time Control of Open Loop Articulated Chains." *Trans. ASME J. Dynam. Syst. Meas. Contr.* 1971; September.

[13] R. H. Taylor. "Planning and Execution of Straight Line Manipulator Trajectories." *IBM J. Res. Dev.* 1979; 23(4).

References for Manipulator Dynamics

[14] J.J. Uicker. *On the Dynamic Analysis of Spatial Linkages Using 4X4 Matrices.* Ph.D. dissertation, Evanston, Ill.: Northwestern University, 1965.

[15] M. Renaud. *Contribution à l'Etude de la Modélisation et de la Commande des Systèmes Mécaniques Articulés.* Toulouse, France: Université Paul Sabatier,

December 1975. Thèse de Docteur Ingénieur.

[16] A. Liegois, W. Khalil, J. M. Dumas, and M. Renaud. "Mathematical and Models of Interconnected Mechanical Systems." Symposium on the Theory and Practice of Robots and Manipulators, Warsaw, Poland, 1976.

[17] Y. Stepanenko and M. Vukobratovic. "Dynamics of Articulated Open-Chain Active Mechanisms." *Math-Biosciences* 1976; 28:137–170.

[18] D. E. Orin et al. "Kinematic and Kinetic Analysis of Open-Chain Linkages Utilizing Newton-Euler Methods." *Math-Biosciences* 1979; 43:107–130.

[19] W. W. Armstrong. "Recursive Solution to the Equations of Motion of an N-Link Manipulator." Proceedings of the 5th World Congress on the Theory of Machines and Mechanisms, Montreal, Quebec, Canada, July 1979.

[20] J. Y. S. Luh, M. W. Walker, R. P. Paul. "On-Line Computational Scheme for Mechanical Manipulators." *Trans. ASME J. Dynam. Syst. Meas. Contr.* 1980.

[21] J. M. Hollerbach. "A Recursive Lagrangian Formulation of Manipulator Dynamics and a Comparative Study of Dynamic Formulation Complexity." *IEEE Trans. Syst. Man Cybernet.* November 1980.

[22] D. B. Silver. "On the Equivalence of Lagrangian and Newton-Euler Dynamics for Manipulators." *Int. J. Robotics Res.* 1982; 1 & 2.

[23] J. M. Hollerbach and G. Sahar. "Wrist-Partitioned Inverse Accelerations and Manipulator Dynamics." Cambridge, Mass.: MIT AI Memo No. 717, April 1983.

[24] T.K. Kanade, P.K. Khosla, and N. Tanaka. "Real-Time Control of the CMU Direct Drive Arm II Using Customized Inverse Dynamics." Proceedings of the 23rd IEEE Conference on Decision and Control, Las Vegas, Nevada, December 1984.

[25] A. Izaguirre and R.P. Paul. "Computation of the Inertial and Gravitational Coefficients of the Dynamic Equations for a Robot Manipulator with a Load." Proceedings of the 1985 International Conference on Robotics and Automation, St. Louis, March 1985:1024–1032.

[26] B. Armstrong, O. Khatib, and J. Burdick. "The Explicit Dynamic Model and Inertial Parameters of the PUMA 560 Arm." Proceedings of the 1986 IEEE International Conference on Robotics and Automation, San Francisco, April 7–11, 1986:510–518.

[27] J.W. Burdick. "An Algorithm for Generation of Efficient Manipulator Dynamic Equations." Proceedings of the 1986 IEEE International Conference on Robotics and Automation, San Francisco, April 7–11, 1986:212–218.

[28] T.R. Kane and D.A. Levinson. "The Use of Kane's Dynamical Equations in Robotics." *Int. J. Robotics Res.* 1983; 2(3):3–20.

[29] M. Renaud. "An Efficient Iterative Analytical Procedure for Obtaining a Robot Manipulator Dynamic Model." First International Symposium of Robotics Research, New Hampshire, August 1983.

References for Manipulator Control

[30] E. Freund. "The Structure of Decoupled Non Linear Systems." *Int. J. Contr.* 1975; 21(3):443–450.

[31] *Idem*. "A Non-Linear Control Concept for Computer-Controlled Manipulators." Proceedings of the 7th ISIR, Tokyo, 1977.

[32] A. K. Bejczy. "Robot Arm Dynamics and Control." JPL NASA Technical Memorandum 33-669, February 1974.

[33] R. A. Lewis. "Autonomous Manipulation on a Robot: Summary of Manipulator Software Functions." JPL Technical Memorandum 33-679, March 1974.

[34] B. R. Markiewicz. "Analysis of the Computed Torque Drive Method and Comparison with Conventional Position Servo for a Computer-Controlled Manipulator." JPL Technical Memorandum 33-601, March 1973.

[35] J. Zabala Iturralde. *Commande des Robots Manipulateurs à partir de la Modélisation de leur Dynamique*. Toulouse, France: Université of Paul Sabatier, July 1978. Thèse de 3eme Cycle.

[36] O. Khatib, M. Llibre, and R. Mampey. "Fonction Décision-Commande d'un Robot Manipulateur." Rapport Scientifique no. 2/7156, DERA-CERT, Toulouse, France, July 1978.

[37] A. Liegois, A. Fournier, and M.J. Aldon. "Model Reference Control of High Velocity Industrial Robots." Proceedings of 1980 JACC, San Francisco, August 1980.

[38] J.Y.S. Luh, M.W. Walker, and R.P. Paul. "Resolved-Acceleration Control of Mechanical Manipulators." *IEEE Trans. Auto. Contr.* 1980; AC-25(3).

[39] S. Arimoto and F. Miyazaki. "Stability and Robustness of PID Feedback Control for Robot Manipulators of Sensory Capability." Third International Symposium of Robotics Research, Gouvieux, France, July 1985.

[40] D. Koditschek. "Adaptive Strategies for the Control of Natural Motion." Proceedings of the 24th Conference on Decision and Control, Fort Lauderdale, Fla., December 1985.

[41] O. Khatib. "Dynamic Control of Manipulators in Operational Space." Sixth IFTOMM Congress on Theory of Machines and Mechanisms, New Delhi, December 15–20, 1983.

[42] J.J. Craig. *Introduction to Robotics: Mechanics and Control*. Reading, Mass.: Addison–Wesley, 1986.

References for General Control, Analysis, Robustness, etc.

[43] M. Spong and M. Vidyasagar. "Robust Nonlinear Control of Robot Manipulators." Proceedings of the 24th Conference on Decision and Control, Fort Lauderdale, Fla., December 1985.

[44] S. Arimoto and F. Miyazaki. "Stability and Robustness of PID Feedback Control for Robot Manipulators of Sensory Capability." Third International Symposium of Robotics Research, Gouvieux, France, July 1985.

[45] J.J. Slotine. "The Robust Control of Robot Manipulators." *Int. J. Robotics Res.*, 1985; 4(2).

[46] I.J. Ha and E. Gilbert. "Robust Tracking in Nonlinear Systems and its Applications to Robotics." IEEE Conference on Decision and Control, Fort Lauderdale, Fla., 1985.

[47] O. Egeland. "On the Robustness of the Computed Torque Technique in Manipulator Control." Proceedings of the IEEE Conference on Robotics and Automation, San Francisco, April 1986.

[48] C. Desoer and M. Vidyasagar. *Feedback Systems: Input-Output Properties.* New York: Academic Press, 1975.

[49] A.M. Lyapunov. "On the General Problem of Stability of Motion." Soviet Union: Kharkov Mathematical Society, 1892 (in Russian).

[50] J. LaSalle and S. Lefschetz. *Stability by Lyapunov's Direct Method with Applications.* New York: Academic Press, 1961.

[51] M. Vidyasagar. *Nonlinear Systems Analysis.* Englewood Cliffs, N.J.: Prentice-Hall, 1978.

[52] K. Atkinson. *An Introduction to Numerical Analysis.* New York: Wiley, 1978.

[53] W. Rudin. *Principles of Mathematical Analysis.* New York: McGraw-Hill, 1976.

[54] H.L. Royden. *Real Analysis.* New York: MacMillan, 1963.

[55] J.J. Craig. "Adaptive Control of Manipulators." Stanford, Calif.: Stanford University, April 1983. Ph.D. thesis proposal.

[56] K. Narendra and A.M. Annaswamy. "Robust Adaptive Control in the Presence of Bounded Disturbances." *IEEE Trans. Auto. Contr.* 1986; AC-31.

[57] S. Boyd and S. Sastry. "On Parameter Convergence in Adaptive Control." *Syst. Contr. Lett.* 1983; 3.

[58] C.E. Rohrs. *Adaptive Control in the Presence of Unmodelled Dynamics.* Cambridge, Mass.: Massachusetts Institute of Technology, November 1982. Ph.D. dissertation.

[59] P.A. Ioannou. *Robustness of Model Reference Adaptive Schemes with Respect to Modelling Errors.* University of Illinois, August 1982. Ph.D. dissertation.

[60] B.B. Peterson and K. Narendra. "Bounded Error Adaptive Control." *IEEE Trans. Auto. Contr.* 1982; AC-27.

[61] M. Corless and G. Leitmann. "Adaptive Control of Systems Containing Uncertain Functions and Unknown Functions with Uncertain Bounds." *J. Optim. Theory Applic.* 1983; 41(1).

[62] G. Goodwin and K.S. Sin. *Adaptive Filtering, Prediction, and Control.* Englewood Cliffs, N.J.: Prentice-Hall, 1984.

[63] S.S. Sastry. "Model-reference adaptive control—Stability, Parameter Convergence, and Robustness." *IMA J. Math. Contr. Info.* 1984.

[64] B.D.O. Anderson and S. Vongpanitlerd. *Network Synthesis: A State Space Approach.* Englewood Cliffs, N.J.: Prentice-Hall, 1973.

[65] Parks. "Liapunov Redesign of Model Reference Adaptive Control Systems." *IEEE Trans. Auto. Contr.* 1966; AC-11(3).

[66] A.P. Morgan and K.S. Narendra. "On the Uniform Asymptotic Stability of Certain Linear Nonautonomous Differential Equations." *SIAM J. Contr. Optim.* 1977; 15.

[67] B.D.O. Anderson. "Exponential Stability of Linear Equations Arising in Adaptive Identification." *IEEE Trans. Auto. Contr.* 1977; AC-22: 83–88.

[68] G. Kreisselmeier. "Adaptive Observers with Exponential Rate of Convergence." *IEEE Trans. Auto. Contr.* 1977; AC-22.

References for Adaptive Control of Manipulators

[69] R.P. Anex and M. Hubbard. "Modeling and Adaptive Control of a Mechanical Manipulator." *ASME J. Dynam. Syst. Meas. Contr.* 1984; 106.

[70] P.G. Backes. *Real Time Control with Adaptive Manipulator Control Schemes.* Purdue University, School of Mechanical Engineering, West Lafayette, Ind.: December 1984. M.S. Thesis.

[71] P.G. Backes, G. Leininger, and C.H. Chung. "Real Time Cartesian Coordinate Hybrid Control of a Puma 560 Manipulator." IEEE Conference on Robotics and Automation, St. Louis, March 1985.

[72] *Idem*. "Joint Self-Tuning with Cartesian Setpoints." 24th IEEE Conference on Decision and Control, Fort Lauderdale, Fla., 1985.

[73] A. Balestrino, G. De Maria, and L. Sciavicco. "Adaptive Control of Robotic Manipulators." AFCET, Congres Automatique, Nantes, France, October 1981.

[74] *Idem*. "An Adaptive Model Following Control for Robotic Manipulators." *ASME J. Dynam. Syst. Meas. Contr.* 1983; 105.

[75] *Idem*. "Adaptive Control of Manipulators in the Task Oriented Space." Proceedings of the 13th ISIR, p13-13 to p13-28, 1983.

[76] R.A. Boie, D. Mitra, W.L. Nelson. "End-Point Sensing and Adaptive Control of a Flexible Robot Arm." 24th IEEE Conference on Decision and Control, Fort Lauderdale, Fla., December 1985.

[77] C. Canudas, K.J. Astrom, and K. Braun. "Adaptive Friction Compensation in DC Motor Drives." IEEE Robotics and Automation Conference, San Francisco, April 1986.

[78] C.T. Cao, "A Simple Adaptive Concept for the Control of an Industrial Robot." *Lecture Notes in Contr. Inf. Sci.* 1980; 24.

[79] Y.K. Choi, M.J. Chung, Z. Bien, and J. Lyou. "An Adaptive Regulation Scheme for Manipulators." International Conference on Advanced Robotics, Tokyo, September 1985.

[80] C.H. Chung and G. Leininger. "Adaptive Self-Tuning Control of Manipulators in Task Coordinate System." International Conference on Robotics, Atlanta,

March 13–15, 1984.

[81] J.J. Craig. "An Adaptive Algorithm for the Computed Torque Method of Manipulator Control." NSF Robotics Workshop on "Intelligent Robots: Achievements and Issues," November 13–14, 1984. Proceedings published March 1985.

[82] J.J. Craig, P. Hsu, S. Sastry. "Adaptive Control of Mechanical Manipulators." IEEE Conference on Robotics and Automation, San Francisco, April 1986.

[83] S. Dubowsky and D.T. DesForges. "The Application of Model-Referenced Adaptive Control to Robotic Manipulators." *ASME J. Dynam. Syst. Meas. Contr.* 1979; 101.

[84] S. Dubowsky. "On the Adaptive Control of Robotic Manipulators: The Discrete Time Case." Proceedings of the Joint Automatic Control Conference, Charlottesville, Va., 1981.

[85] S. Dubowsky and R. Kornbluh. "On the Development of High Performance Adaptive Control Algorithms for Robotic Manipulators." 2nd International Symposium on Robotics Research, Kyoto, Japan, 1984.

[86] H. Durrant-Whyte. "Practical Adaptive Control of Actuated Spatial Mechanisms." IEEE Conference on Robotics and Automation, St. Louis, March 1985.

[87] H. Elliot, T. Depkovich, J. Kelly, and B. Draper. "Nonlinear Adaptive Control of Mechanical Linkage Systems with Application to Robotics." American Control Conference, San Diego, June 1984.

[88] R. M. Goor. "Continuous Time Adaptive Feedforward Control: Stability and Simulations." General Motors Research Publication GMR-4105, July 1982.

[89] H. Hemami and Y. F. Zheng. "An Adaptive Positional Control Scheme for Robotic Systems." 24th IEEE Conference on Decision and Control, Fort Lauderdale, Fla., 1985.

[90] R. Horowitz and M. Tomizuka. "An Adaptive Control Scheme for Mechanical Manipulators—Compensation of Nonlinearity and Decoupling Control." ASME Paper no. 80-WA/DSC-6, 1980.

[91] R. Horowitz and M. Tomizuka. "Discrete Time Model Reference Adaptive Control of Mechanical Manipulators." *Comput. Eng.* 1982; 2:107–112.

[92] R. Horowitz and M. Tomizuka, personal communication, 1986.

[93] T. C. Hsia. "Adaptive Control of Robot Manipulators—A Review." IEEE Conference on Robotics and Automation, San Francisco, 1986.

[94] B. K. Kim and K. G. Shin. "An Adaptive Model Following Control of Industrial Manipulators." *IEEE Trans. Aerosp. Elect. Syst.* Nov. 1983; AES-19:805–814.

[95] D. E. Koditschek. "Adaptive Strategies for the Control of Natural Motion." Proceedings of the 24th Conference on Decision and Control, Fort Lauderdale, Fla., December 1985.

[96] A. J. Koivo and R. P. Paul. "Manipulators with Self-Tuning Controllers." IEEE Conference on Cybernetics and Society, Boston, 1980.

[97] A.J. Koivo et al. "Control of Robotic Manipulator with Adaptive Controller." Proceedings of the IEEE Conference on Decision and Control, San Diego, 1981.

[98] A.J. Koivo, Q. Zhang, and T.H. Guo. "Industrial Manipulator Control Using Kalman Filter and Adaptive Controller." Proceedings of the 1982 IFAC Symposium on Identification and Systematic Parameter Estimation, Arlington, Va., June 7–11, 1982.

[99] A.J. Koivo and T. Guo. "Adaptive Linear Controller for Robotic Manipulators." *IEEE Trans. Auto. Contr.* 1983; AC-28(2).

[100] A.J. Koivo. "Robotic Manipulator with Self-Tuning Controller Operating in Cartesian Coordinates." Proceedings of the Third Yale Workshop on Applications of Adaptive System Theory, June 1983:174–178.

[101] A.J. Koivo, R. Lewczyk, T.H. Chiu. "Adaptive Path Control of a Manipulator with Visual Information." IEEE Conference on Robotics, Atlanta, March 1984.

[102] A.J. Koivo. "Adaptive Position-Velocity-Force Control of Two Manipulators." 24th IEEE Conference on Decision and Control, Fort Lauderdale, Fla., 1985.

[103] A.J. Koivo. "Force-Velocity Control with Self-Tuning for Robotic Manipulators." IEEE Conference on Robotics and Automation, San Francisco, April 1986.

[104] H.N. Koivo and J. Sorvari. "On Line Tuning of Multivariable PID-Controller for Robot Manipulators." 24th IEEE Conference on Decision and Control, Fort Lauderdale, Fla., 1985.

[105] K. Kubo and T. Ohmae. "Adaptive Trajectory Control of Industrial Robots." 15th ISIR, Tokyo, Japan, September 1985.

[106] M. LeBorgne, J.M. Ibarra, B. Espiau. "Adaptive Control of High Velocity Manipulators." 11th ISIR, Tokyo, Japan, 1981.

[107] C.S.G. Lee and M.J. Chung. "An Adaptive Control Strategy for Computer-Based Manipulator." 21st Conference on Decision and Control, Orlando, Fla., 1982:95–100.

[108] C.S.G. Lee and B.H. Lee. "Resolved Motion Adaptive Control for Mechanical Manipulators." Proceedings of the Third Yale Workshop on Applications of Adaptive System Theory, June 1983:190–196.

[109] C.S.G. Lee, M.J. Chung, and B.H. Lee. "Adaptive Control for Robot Manipulators in Joint and Cartesian Coordinates." IEEE Conference on Robotics, Atlanta, March 1984.

[110] *Idem.* "An Approach of Adaptive Control for Robotic Manipulators." *J. Robot. Syst.* 1984; 1.

[111] C.S.G. Lee and M.J. Chung. "An Adaptive Control Strategy for Mechanical Manipulators." *IEEE Trans. Auto. Contr.* 1984; AC-29(9).

[112] S. Lee. "Reduced Model Inverse Control with Adaptive Modelling-Error Compensation for Robot Arms." 24th IEEE Conference on Decision and Control, Fort Lauderdale, Fla., December 1985.

[113] G. Leininger and S. Wang. "Pole Placement Self-Tuning Control of Manipulators." IFAC Symposium on Computer Aided Design of Multivariate Technological Systems, West Lafayette, Ind., September 15–17, 1982.

[114] G. Leininger. "Self-Tuning Control of Manipulators." International Symposium on Advanced Software in Robotics, Liege, Belgium, May 1983.

[115] G. Leininger. "Adaptive Control of Manipulators Using Self-Tuning Methods." 1st ISRR, New Hampshire, August 29–September 2, 1983.

[116] K. Y. Lim and M. Eslami. "Adaptive Controller Designs for Robot Manipulator Systems Using Lyapunov Direct Method." *IEEE Trans. Auto. Contr.* 1985; 30(12).

[117] K. Y. Lim and M. Eslami. "Improvement of Adaptive Controller Designs for Robot Manipulator Systems Yielding Reduced Cartesian Error." Proceedings of the 24th IEEE Conference on Decision and Control, December 1985.

[118] K. Y. Lim and M. Eslami. "Robust Adaptive Controller Designs for Robot Manipulator Systems." IEEE Conference on Robotics and Automation, San Francisco, April 1986.

[119] M. H. Liu. "An Adaptive Control Scheme for Robotic Manipulators." 15th ISIR, Tokyo, September 1985.

[120] Z. Ma and C. S. G. Lee. "Adaptive Perturbation Control with Compensation for Disturbances for Robot Manipulators." 24th IEEE Conference on Decision and Control, Fort Lauderdale, Fla., December 1985.

[121] V. A. Malyshev and A. V. Timofeyev. "Manipulator Dynamics and Adaptive Control." *Auto. Remote Contr.* 1981; 42(8):1069–1076.

[122] D. R. Meldrum and M. J. Balas. "Direct Adaptive Control of Flexible Remote

Manipulator Arm." ASME Winter Annual Meeting, Miami, November 1985.

[123] W.L. Nelson and D. Mitra. "Load Estimation and Load-Adaptive Optimal Control for a Flexible Robot Arm." IEEE Conference on Robotics and Automation, San Francisco, April 1986.

[124] C.P. Neuman and H. Stone. "MRAC Control of Robotic Manipulators." Proceedings of the Third Yale Workshop on Adaptive System Theory, June 1983.

[125] F. Nicolo and J. Katende. "A Robust MRAC for Industrial Robots." 2nd IASTED International Symposium on Robotics and Automation, Lugano, Switzerland, 1983.

[126] S. Nicosia and P. Tomei. "Model Reference Adaptive Control Algorithms for Industrial Robots." *Automatica*. 1984; 20(5):635–644.

[127] H. Seraji. "Adaptive Control of Robotic Manipulators." JPL Engineering Memorandum 347-182, January 1986.

[128] J.J. Slotine and J.A. Coetsee. "Adaptive Sliding Controller Synthesis for Non-Linear Systems." *Int. J. Contr.* 1986.

[129] J.J. Slotine. "On Modeling and Adaptation in Robot Control." Proceedings of the IEEE Conference on Robotics and Automation, San Francisco, April 1986.

[130] S.N. Singh. "Adaptive Model Following Control of Nonlinear Robotic Systems." *IEEE Trans. Auto. Contr.* 1985; AC-30(11).

[131] D.P. Stoten. "Discrete Control of a Manipulator Arm." *Optim. Contr. Applic. Methods*. 1982; 3:423–433.

[132] M. K. Sundareshan and M. A. Koenig. "Decentralized Model Reference Adaptive Control of Robotic Manipulators." American Control Conference, 1985:44–49.

[133] A. V. Timofeyev and J. V. Ekalo. "Stability and Stabilization of Programmed Motion of Robot-Manipulators." *Automatika and Telemechanika*. 1976; 10:148–156. (in Russian).

[134] P. Tomei, S. Nicosia, and A. Ficola. "An Approach to the Adaptive Control of Robots Having Elastic Joints." IEEE Conference on Robotics and Automation, San Francisco, April 1986.

[135] M. Tomizuka, R. Horowitz, and Y. D. Landau. "On the Use of Model Reference Adaptive Control Techniques for Mechanical Manipulators." 2nd IASTED Symposium on Identification, Control and Robotics, Davos, Switzerland, March 1982.

[136] M. Tomizuka and R. Horowitz. "Model Reference Adaptive Control of Mechanical Manipulators." IFAC Adaptive Systems in Control and Signal Processing, San Francisco, 1983.

[137] R. G. Walters. "Application of a Self Tuning Pole-Placement Regulator to an Industrial Manipulator." IEEE Conference on Decision and Control, 1982:323.

References for Learning Control of Manipulators

[138] M. Uchiyama. "Formation of High Speed Motion Pattern of Mechanical Arm by Trial." *Trans. Soc. Instr. Contr. Eng.* 1978; 19(5):706–712.

[139] J.J. Craig. "Adaptive Control of Manipulators Through Repeated Trials."

General Motors Research Report GMR-4530, December 1983, and Proceedings of the American Control Conference, San Diego, June 1984.

[140] S. Arimoto. "Mathematical Theory of Learning with Applications to Robot Control." Proceedings of the Fourth Yale Workshop on Applications of Adaptive Systems Theory, May 1985.

[141] S. Arimoto, S. Kawamura, and F. Miyazaki. "Bettering Operation of Robots by Learning." *J. Robot. Syst.* 1984; 1&2:123–140.

[142] S. Arimoto, S. Kawamura, and F. Miyazaki. "Bettering Operation of Dynamic Systems by Learning: A New Control Theory for Servomechanisms and Mechatronics Systems." Proceedings of the 23rd Conference on Decision and Control, Las Vegas, December 1984.

[143] S. Arimoto, S. Kawamura, F. Miyazaki, and S. Tamaki. "Learning Control Theory for Dynamical Systems." Proceedings of the 24th Conference on Decision and Control, Fort Lauderdale, Fla., December 1985.

[144] S. Hara, T. Omata, and M. Nakano. "Synthesis of Repetitive Control Systems and Its Applications." Proceedings of the 24th Conference on Decision and Control, Fort Lauderdale, Fla., December 1985.

[145] S. Kawamura, F. Miyazaki, and S. Arimoto. "Applications of Learning Method for Dynamic Control of Robot Manipulators." Proceedings of the 24th Conference on Decision and Control, Fort Lauderdale, Fla., December 1985.

[146] *Idem.* "Iterative Learning Control for Robotic Systems." Proceedings of the IECON 84, Tokyo, Japan, October 1984.

[147] T. Mita and E. Kato. "Iterative Control and its Application to Motion Control of Robot Arm—A Direct Approach to Servo Problems." Proceedings of the 24th Conference on Decision and Control, Fort Lauderdale, Fla., December 1985.

[148] M. Togai and O. Yamano. "Analysis and Design of an Optimal Learning Control Scheme for Industrial Robots: A Discrete Time Approach." Proceedings of the 24th Conference on Decision and Control, Fort Lauderdale, Fla., December 1985.

[149] S. H. Wang. "Computed Reference Error Adjustment Technique (CREATE) for the Control of Robot Manipulators." 22nd Annual Allerton Conference on Communication, Control, and Computing, October 1984.

[150] S. H Wang and I. Horowitz. "CREATE—A New Adaptive Technique." Proceedings of the 19th Annual Conference on Information Sciences and Systems, March 1985.

[151] C. G. Atkeson and J. McIntyre. "Robot Trajectory Learning Through Practice." Proceedings of the IEEE Conference on Robotics and Automation, San Francisco, 1986.

INDEX